T0295083

Inventory Planning with Forecasting Expenditure

Inventory Planning with Forecasting Expenditure

Sanjay Sharma

CRC Press
Taylor & Francis Group
Boca Raton London New York

CRC Press is an imprint of the
Taylor & Francis Group, an **informa** business

First edition published 2022
by CRC Press
6000 Broken Sound Parkway NW, Suite 300, Boca Raton, FL 33487-2742

and by CRC Press
4 Park Square, Milton Park, Abingdon, Oxon, OX14 4RN

First edition published by CRC Press 2022

CRC Press is an imprint of Taylor & Francis Group, LLC

ISBN: 978-1-032-20929-6 (hbk)
ISBN: 978-1-032-21202-9 (pbk)
ISBN: 978-1-003-26725-6 (ebk)

DOI: 10.1201/9781003267256

Typeset in Times
by MPS Limited, Dehradun

Contents

Preface

In an industrial or business case, purchase or procurement is a significant function. Usually a procurement plan is prepared on the basis of a certain prediction of consumption pattern or demand. However when this plan is implemented, then the desired benefit is obtained corresponding to the forecast accuracy. In the available literature, the forecasting accuracy is discussed a lot. A need is established to link the forecasting accuracy with a forecasting expenditure. After an explicit inclusion of the forecasting expenditure, the inventory planning for procurement/production is described in the present book.

This book is expected to be useful for UG/PG students in engineering and management and also has good potential for elective/supplementary core courses.

Sanjay Sharma
Mumbai, India

Author

Sanjay Sharma is a professor at the National Institute of Industrial Engineering (NITIE), Mumbai, India. He is an operations and supply chain management educator and researcher with more than three decades of experience, including industrial, managerial, teaching and training, consultancy, and research. Professor Sharma has many awards and honors to his credit. He has published eight books and papers in various journals such as *European Journal of Operational Research*, *International Journal of Production Economics*, *Computers & Operations Research*, *International Journal of Advanced Manufacturing Technology*, *Journal of the Operational Research Society*, and *Computers and Industrial Engineering*. He is also on the editorial board of several journals, including the *International Journal of Logistics Management*.

1 Introduction

In an industrial or business case, purchase or procurement is a significant function. Usually a procurement plan is prepared on the basis of a certain prediction of consumption pattern or demand. However, when this plan is implemented, then the desired benefit is obtained corresponding to the forecast accuracy. In the available literature, forecasting accuracy is discussed a lot. A need is established to link the forecasting accuracy with a forecasting expenditure. After an explicit inclusion of the forecasting expenditure, the inventory planning for procurement/production is described in this book.

In this introductory chapter, a problem description is extended to incorporate the forecasting expenditure. The potential benefit is also mentioned briefly so that it can be incorporated in subsequent related chapters. However, first a planned schedule is explained that is relevant for a procurement as well as a production scenario.

1.1 PLANNED SCHEDULE

A schedule refers to the periods in which a certain quantity is either procured or produced.

Consider the following requirement for purchase:

Period:	1	2	3	4	5	6	7	8	9	10	11	12
Demand:	50	50	50	50	50	50	50	50	50	50	50	50

In order to meet the requirement, a procurement can happen in multiple ways, such as shown below:

(a)

Period:	1	2	3	4	5	6	7	8	9	10	11	12
Demand:	50	50	50	50	50	50	50	50	50	50	50	50
Purchase quantity:	50	50	50	50	50	50	50	50	50	50	50	50

That is, the respective quantity is purchased or ordered in every period to satisfy the demand or requirement.

(b)

Period:	1	2	3	4	5	6	7	8	9	10	11	12
Demand:	50	50	50	50	50	50	50	50	50	50	50	50
Purchase quantity:	100	0	100	0	100	0	100	0	100	0	100	0

That is, the demand is purchased or ordered for the two periods. Thus, the six orders are placed for a quantity of 100 units.

DOI: 10.1201/9781003267256-1

(c)

Period:	1	2	3	4	5	6	7	8	9	10	11	12
Demand:	50	50	50	50	50	50	50	50	50	50	50	50
Purchase quantity:	150	0	0	150	0	0	150	0	0	150	0	0

That is, the demand is purchased or ordered for the three periods. Thus, the four orders are placed for a quantity of 150 units.

(d)

Period:	1	2	3	4	5	6	7	8	9	10	11	12
Demand:	50	50	50	50	50	50	50	50	50	50	50	50
Purchase quantity:	200	0	0	0	200	0	0	0	200	0	0	0

That is, the demand is purchased or ordered for the four periods. Thus, the three orders are placed for a quantity of 200 units.

(e)

Period:	1	2	3	4	5	6	7	8	9	10	11	12
Demand:	50	50	50	50	50	50	50	50	50	50	50	50
Purchase quantity:	300	0	0	0	0	0	300	0	0	0	0	0

That is, the demand is purchased or ordered for six periods. Thus, the two orders are placed for a quantity of 300 units.

(f)

Period:	1	2	3	4	5	6	7	8	9	10	11	12
Demand:	50	50	50	50	50	50	50	50	50	50	50	50
Purchase quantity:	600	0	0	0	0	0	0	0	0	0	0	0

That is, the demand for all 12 periods is purchased or ordered at once. Thus, only one order is placed for a quantity of 600 units.

As the number of orders becomes lower, the ordering quantity increases. Thus, the efforts pertaining to the order placement process become lower, resulting in a lower ordering cost. However, as the ordering quantity increases, inventory holding cost increases. Thus, the planned procurement schedule should take into consideration the ordering cost as well as an inventory carrying or holding cost.

In a manufacturing context, the facility or a group of facilities is set up for production activity. Efforts are needed to set up the facility in order to produce a certain number of products, i.e., the batch quantity. Efforts concerning a setup and the related costs are estimated to obtain a setup cost. Thus, a facility setup cost becomes relevant instead of an ordering cost. Consider the following requirement for production:

Period:	1	2	3	4	5	6	7	8	9	10	11	12
Demand:	75	75	75	75	75	75	75	75	75	75	75	75

In order to meet the requirement, a production can happen in multiple ways, such as shown below:

(a)

Period:	1	2	3	4	5	6	7	8	9	10	11	12
Demand:	75	75	75	75	75	75	75	75	75	75	75	75
Batch quantity:	75	75	75	75	75	75	75	75	75	75	75	75

That is, the respective production batch quantity is manufactured in every period to satisfy the demand or requirement.

(b)

Period:	1	2	3	4	5	6	7	8	9	10	11	12
Demand:	75	75	75	75	75	75	75	75	75	75	75	75
Batch quantity:	150	0	150	0	150	0	150	0	150	0	150	0

That is, the demand for two periods is produced in a facility setup. Thus, six setups are needed for a batch quantity of 150 units.

(c)

Period:	1	2	3	4	5	6	7	8	9	10	11	12
Demand:	75	75	75	75	75	75	75	75	75	75	75	75
Batch quantity:	225	0	0	225	0	0	225	0	0	225	0	0

That is, the demand for three periods is produced in a facility setup. Thus, the four setups are needed for a batch quantity of 225 units.

(d)

Period:	1	2	3	4	5	6	7	8	9	10	11	12
Demand:	75	75	75	75	75	75	75	75	75	75	75	75
Batch quantity:	300	0	0	0	300	0	0	0	300	0	0	0

That is, the demand for four periods is produced in a facility setup. Thus, the three setups are needed for a batch quantity of 300 units.

(e)

Period:	1	2	3	4	5	6	7	8	9	10	11	12
Demand:	75	75	75	75	75	75	75	75	75	75	75	75
Batch quantity:	450	0	0	0	0	0	450	0	0	0	0	0

That is, the demand for six periods is produced in a facility setup. Thus, the two setups are needed for a batch quantity is 450 units.

(f)

Period:	1	2	3	4	5	6	7	8	9	10	11	12
Demand:	75	75	75	75	75	75	75	75	75	75	75	75
Batch quantity:	900	0	0	0	0	0	0	0	0	0	0	0

That is, the demand for all 12 periods is produced at once. Thus, only one facility setup is needed for a quantity of 900 units.

As the number of facility setups lowers, the production batch quantity increases. Thus, the efforts pertaining to the manufacturing setup become lower, resulting in a lower setup cost. However, as the batch quantity increases, inventory holding cost increases. Thus, the planned production schedule should take into consideration the setup cost as well as an inventory carrying or holding cost.

1.2   PROBLEM DESCRIPTION

Conventionally, the schedule is planned considering the cost components as follows:

    i. Inventory carrying cost
    ii. Ordering cost or the facility setup cost

In the procurement situation, an ordering cost is relevant, whereas in the production case, the facility setup cost is incorporated. Inventory carrying or holding cost is relevant in both cases, i.e., procurement and production.

Most of the businesses depend heavily on the demand forecast. However, the existing scope is confined to the forecasting accuracy only. For example, the demand in a certain period is predicted as 200 units and accordingly the arrangements have been made, but the actual demand may vary. It might be as follows:

    i. Greater than 200
    ii. Less than 200

If it is greater than 200, then a portion of demand is not met. Extra handling and space management is needed if it is less than 200 units. The less the deviation on both sides, the better the forecasting accuracy. The businesses usually strive hard to lower the deviation or improve the forecasting accuracy. However, the forecasting efforts and an associated expenditure is usually ignored. An approach for justification of such an expenditure is shown in Figure 1.1.

In cases where certain businesses depend heavily on the demand forecast, then a thorough study of the deviation from the predicted demand would become necessary. A need for improvement is established when this deviation is significant, in the opinion of a concerned management. However, the intended improvement certainly requires an additional forecasting effort. As the associated efforts include resources, such as time and also the human resources among others, there is strong justification for an explicit inclusion of a forecasting expenditure in the decision-making approach. Presently, this problem is described and handled rigorously in the context of procurement and production activities.

1.3   FORECASTING EXPENDITURE

An expenditure toward forecasting needs to be assessed. This may include time and resources required in addition to the coordination efforts. Inclusion of the estimated forecasting expenditure can be done in a variety of ways such as for the whole planning horizon or on an annual basis. However it is much more convenient when

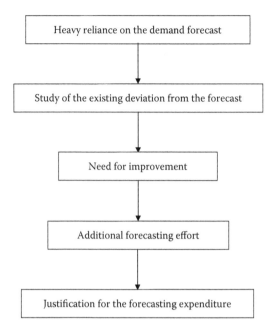

FIGURE 1.1 Approach for justifying the forecasting expenditure.

this is included in the cyclic cost. An inventory cycle may relate to buying or making functions.

1.3.1 FOR BUYING

In order to buy an item, assume that the annual demand is 600 units and it is procured in quantities such as 200 units. Thus, there are three inventory cycles in a year for buying the item. In each cycle, the ordering cost and certain forecasting expenditure are incurred. An annual cost concerning such cost components can be expressed as:

$$= \frac{D}{Q}(C + M) \tag{1.1}$$

where

D = Annual demand
Q = Ordering quantity in each cycle
C = Ordering cost
M = Estimated forecasting expenditure per cycle

As the average inventory throughout the year is $\frac{Q}{2}$, an annual cost related to the inventory carrying can be expressed as:

$$= \frac{Q}{2} \cdot I \tag{1.2}$$

where

I = Annual inventory holding cost per unit

A total related cost (E) can be expressed after adding Eqs. (1.1) and (1.2):

$$E = \frac{D}{Q}(C + M) + \frac{Q}{2} \cdot I \tag{1.3}$$

In order to minimize the total cost, differentiate with respect to Q and equate to zero. Thus:

$$\frac{D(C + M)}{Q^2} = \frac{I}{2}$$

An optimal ordering quantity in each cycle can be expressed as:

$$Q = \sqrt{\frac{2D(C + M)}{I}} \tag{1.4}$$

Substituting this optimal value of Q in Eq. (1.3), a minimized total cost can be expressed as:

$$E = \sqrt{2DI(C + M)} \tag{1.5}$$

Example 1.1: Let:

D = 600
C = ₹100
M = ₹100
I = ₹6

A minimized total cost can be obtained from Eq. (1.5) as:

E = ₹1200

Since forecasting expenditure M is introduced, it can be analyzed furthermore. The impact of its variation, i.e., the increase and decrease, can be observed. Let:

h = Percentage variation in M
b = Percentage variation in E

With an increase in M:

$$E\left(1 + \frac{b}{100}\right) = \sqrt{2DI\{C + M(1 + h/100)\}}$$

Substituting Eq. (1.5):

$$1 + \frac{b}{100} = \frac{\sqrt{2DI\{C + M(1 + h/100)\}}}{\sqrt{2DI(C + M)}}$$

or

$$\frac{b}{100} = \sqrt{\frac{C + M(1 + h/100)}{(C + M)}} - 1$$

or

$$b = \left[\sqrt{\frac{C + M(1 + h/100)}{(C + M)}} - 1\right] * 100 \qquad (1.6)$$

Table 1.1 shows the percentage increase in E with respect to the percentage increase in M.

With a decrease in M:

$$E\left(1 - \frac{b}{100}\right) = \sqrt{2DI\{C + M(1 - h/100)\}}$$

Substituting Eq. (1.5):

$$1 - \frac{b}{100} = \frac{\sqrt{2DI\{C + M(1 - h/100)\}}}{\sqrt{2DI(C + M)}}$$

or

$$\frac{b}{100} = 1 - \sqrt{\frac{C + M(1 - h/100)}{(C + M)}}$$

or

$$b = \left[1 - \sqrt{\frac{C + M(1 - h/100)}{(C + M)}}\right] * 100 \qquad (1.7)$$

TABLE 1.1
Value of b with Increase in M

$h =$	5	10	15	20
$b =$	1.24	2.47	3.68	4.88

TABLE 1.2

Value of b with Decrease in M

$h =$	5	10	15	20
$b =$	1.26	2.53	3.82	5.13

Table 1.2 shows the percentage decrease in E with respect to the percentage decrease in M.

Values of b are observed to be higher in comparison with the previous situation. Analytically, also, it can be shown considering Eqs. (1.6) and (1.7) as:

$$1 - \sqrt{\frac{C + M(1 - h/100)}{(C + M)}} > \sqrt{\frac{C + M(1 + h/100)}{(C + M)}} - 1$$

or

$$2 > \sqrt{\frac{C + M(1 + h/100)}{(C + M)}} + \sqrt{\frac{C + M(1 - h/100)}{(C + M)}}$$

or

$$4(C + M)) > C + M(1 + h/100) + C + M(1 - h/100)$$
$$+ 2\sqrt{\{C + M(1 + h/100)\}\{C + M(1 - h/100)\}}$$

or

$$4(C + M) > 2C + 2M$$
$$+ 2\sqrt{C^2 + CM(1 - h/100) + CM(1 + h/100) + M^2\{1 - (h/100)^2\}}$$

or

$$2(C + M) > (C + M) + \sqrt{C^2 + 2CM + M^2 - (Mh/100)^2}$$

or

$$(C + M) > \sqrt{(C + M)^2 - (Mh/100)^2}$$

or

$$(C + M)^2 > (C + M)^2 - (Mh/100)^2$$

Thus, it is true.

Example 1.2: Usually, when the total cost increases, the management likes to find ways so that this can be brought to the previous level. Currently the forecasting expenditure, M, is focused; consider an increase of 10%. Refer to Example 1.1; M is increased from ₹100 to ₹110, resulting in an increase in E. In the case where it is possible to reduce the ordering cost, C, specific efforts might be directed toward this aim. A reduced value of C can be obtained with the help of Eq. (1.5).

$$1200 = \sqrt{2X600X6X(C + 110)}$$

or C = ₹90.

That is, a 10% reduction in C. In order to derive the % reduction in C, let:
  h = Percentage increase in M
  b = Percentage reduction in C

$$\sqrt{2DI\{C(1 - b/100) + M(1 + h/100)\}} = \sqrt{2DI(C + M)}$$

or

$$C(1 - b/100) + M(1 + h/100) = C + M$$

or

$$C - C(1 - b/100) = M(1 + h/100) - M$$

or

$$Cb/100 = Mh/100$$

or

$$Cb = Mh$$

or

$$b = \frac{Mh}{C} \tag{1.8}$$

In order to achieve a similar objective, another effort can be directed towards a reduction in the holding cost, I. Let:

 h = Percentage increase in M
 b = Percentage reduction in I

$$\sqrt{2DI\,(1 - b/100)\{C + M\,(1 + h/100)\}} = \sqrt{2DI\,(C + M)}$$

or

$$(1 - b/100)\{C + M\,(1 + h/100)\} = C + M$$

or

$$C + M\,(1 + h/100) - (b/100)\{C + M\,(1 + h/100)\} = C + M$$

or

$$Mh/100 = (b/100)\{C + M\,(1 + h/100)\}$$

or

$$b = \frac{Mh}{C + M\,(1 + h/100)} \qquad (1.9)$$

For $h = 10$, $b = 4.76$.

   For attainment of similar aim, either of the efforts can be planned, such as:

 i. 10% reduction in C
 ii. 4.76% reduction in I

However, percentage reduction in I is relatively lower because (using Eqs. (1.8) and (1.9)):

$$\frac{Mh}{C + M\,(1 + h/100)} < \frac{Mh}{C}$$

or

$$C + M\,(1 + h/100) > C$$

Furthermore a feasibility of respective reduction in relevant parameter needs to be examined.

   In case where it is not possible to reduce the value of C to the extent of 10%, a combination of both kinds of efforts can be examined, i.e., the reduction in C as well as I. Let:

h = Percentage increase in M
b = Percentage reduction in C
k = Percentage reduction in I

$$\sqrt{2DI(1 - k/100)\{C(1 - b/100) + M(1 + h/100)\}} = \sqrt{2DI(C + M)}$$

or

$$(1 - k/100)\{C(1 - b/100) + M(1 + h/100)\} = C + M$$

or

$$C(1 - b/100) + M(1 + h/100) - (k/100)\{C(1 - b/100) + M(1 + h/100)\}$$
$$= C + M$$

or

$$(Cb/100) + (k/100)\{C(1 - b/100) + M(1 + h/100)\} = (Mh/100)$$

or

$$Cb + k\{C(1 - b/100) + M(1 + h/100)\} = Mh$$

or

$$k = \frac{Mh - Cb}{C(1 - b/100) + M(1 + h/100)}$$

For $h = 10$, Table 1.3 shows various combinations of b and k.

1.3.2 FOR MAKING

In order to make an item, production rate is an important parameter. Inventory buildup happens during the production time in an overall cycle time. For example, if the annual production rate is 1,200 units and production quantity in a cycle is 300 units, then the production time in each cycle is:

TABLE 1.3
Percentage Reduction in C and I

$b =$	2	4	6	8
$k =$	3.85	2.91	1.96	0.99

$$\frac{300}{1200} = 0.25 \text{ year}$$

Let the batch quantity in each cycle be Q and the production rate is P, then the production time is:

$$\frac{Q}{P}$$

During the production time, the inventory buildup also happens at the rate, $(P - D)$ considering the demand rate as D. Thus, the production time can also be expressed as:

$$\frac{V}{(P - D)}$$

where V = Maximum inventory during the cycle.
  Therefore:

$$\frac{V}{(P - D)} = \frac{Q}{P}$$

or

$$V = Q(1 - D/P) \qquad\qquad (1.10)$$

As the average inventory is $\frac{V}{2}$,

  the annual inventory carrying cost = $\frac{V}{2} \cdot I$

$$= \frac{QI(1 - D/P)}{2} \qquad\qquad (1.11)$$

with substitution of Eq. (1.10).
  Since there are $\frac{D}{Q}$ production cycles in a year, the annual facility setup and forecasting expenditure can be given as:

$$= \frac{D}{Q}(C + M) \qquad\qquad (1.12)$$

where
  C = Facility setup cost
  M = Forecasting expenditure per cycle

Adding Eqs. (1.11) and (1.12), the related total cost can be expressed as:

$$E = \frac{QI(1 - D/P)}{2} + \frac{D}{Q}(C + M) \qquad (1.13)$$

In order to minimize the total cost, differentiate with respect to Q and equate to zero. Thus:

$$\frac{D(C + M)}{Q^2} = \frac{I(1 - D/P)}{2}$$

An optimal production batch quantity in each cycle can be expressed as:

$$Q = \sqrt{\frac{2D(C + M)}{I(1 - D/P)}} \qquad (1.14)$$

Substituting this optimal value of Q in Eq. (1.13), a minimized total cost can be expressed as:

$$E = \sqrt{2DI(C + M)(1 - D/P)} \qquad (1.15)$$

Example 1.3: Let:

D = 900
P = 1,200
C = ₹250
M = ₹200
I = ₹36
A minimized total cost can be obtained from Eq. (1.15) as:
E = ₹2700

Since a forecasting expenditure M is introduced, it can be analyzed furthermore. Usually when the total cost increases, the management likes to find ways so that this can be brought to the previous level. Consider an increase of 10% in M, i.e., it is increased from ₹200 to ₹220, resulting in an increase in E. In case where it is possible to reduce the facility setup cost, C, specific efforts might be directed toward this aim. A reduced value of C can be obtained with the help of Eq. (1.15).

$$2700 = \sqrt{2 X 900 X 36 X (C + 220)(1 - 900/1200)}$$

or

C = ₹230
That is, 8% reduction in C.

Let:

h = Percentage increase in M

b = Percentage reduction in C

$$\sqrt{2DI\{C(1 - b/100) + M(1 + h/100)\}(1 - D/P)} = \sqrt{2DI(C + M)(1 - D/P)}$$

or

$$b = \frac{Mh}{C} \tag{1.16}$$

Another option can be the reduction in production rate.

$$2700 = \sqrt{2X900X36X(250 + 220)(1 - 900/P)}$$

or

$$P = 1183.22$$

That is, 1.40% reduction in P.

Let:

h = Percentage increase in M

b = Percentage reduction in P

$$\sqrt{2DI\{C + M(1 + h/100)\}\{1 - D/P(1 - b/100)\}} = \sqrt{2DI(C + M)(1 - D/P)}$$

or

$$\{C + M(1 + h/100)\}\{1 - D/P(1 - b/100)\} = (C + M)(1 - D/P)$$

or

$$C + M(1 + h/100) - \{D/P(1 - b/100)\}\{C + M(1 + h/100)\} = C + M \\ - (C + M)(D/P)$$

or

$$(Mh/100) + (C + M)(D/P) = \{D/P(1 - b/100)\}\{C + M(1 + h/100)\}$$

or

$$\frac{D}{P(1 - b/100)} = \frac{(Mh/100) + (C + M)(D/P)}{C + M(1 + h/100)}$$

or

$$\frac{P(1 - b/100)}{D} = \frac{C + M(1 + h/100)}{(Mh/100) + (C + M)(D/P)}$$

or

$$1 - \frac{b}{100} = \frac{D\{C + M(1 + h/100)\}}{P\{(Mh/100) + (C + M)(D/P)\}}$$

or

$$1 - \frac{b}{100} = \frac{D\{C + M(1 + h/100)\}}{(PMh/100) + (C + M)D}$$

or

$$\frac{b}{100} = \frac{(PMh/100) + (C + M)D - DC - MD(1 + h/100)}{(PMh/100) + (C + M)D}$$

or

$$\frac{b}{100} = \frac{(PMh/100) - (MDh/100)}{(PMh/100) + (C + M)D}$$

or

$$b = \frac{PMh - MDh}{(PMh/100) + (C + M)D}$$

or

$$b = \frac{Mh(P - D)}{(PMh/100) + (C + M)D} \qquad (1.17)$$

For attainment of similar aim, either of the efforts can be planned such as:

  i. 8% reduction in C
 ii. 1.40% reduction in P

However, % reduction in P is relatively lower because (using Eqs. (1.16) and (1.17)):

$$\frac{Mh(P - D)}{(PMh/100) + (C + M)D} < \frac{Mh}{C}$$

or

$$\frac{C(P - D)}{(PMh/100) + (C + M)D} < 1$$

or

$$C(P - D) < (PMh/100) + (C + M)D$$

or

$$(PMh/100) + (C + M)D - C(P - D) > 0$$

or

$$(PMh/100) + CD + MD - CP + CD > 0$$

or

$$(PMh/100) + MD + 2CD > CP$$

For all practical purposes, this is true because the condition is easily satisfied even if P is as high as equivalent to 2-D.

In order to examine a reduction in both the parameters, i.e., C and P, let:

h = Percentage increase in M
b = Percentage reduction in C
k = Percentage reduction in P
Now:

$$\sqrt{2DI\{C(1 - b/100) + M(1 + h/100)\}\{1 - D/P(1 - k/100)\}}$$
$$= \sqrt{2DI(C + M)(1 - D/P)}$$

or

$$\{C(1 - b/100) + M(1 + h/100)\}\{1 - D/P(1 - k/100)\} = (C + M)(1 - D/P)$$

or

$$1 - D/P(1 - k/100) = \frac{(C + M)(1 - D/P)}{C(1 - b/100) + M(1 + h/100)}$$

or

$$\frac{D}{P(1 - k/100)} = \frac{C(1 - b/100) + M(1 + h/100) - (C + M)(1 - D/P)}{C(1 - b/100) + M(1 + h/100)}$$

or

$$\frac{P(1 - k/100)}{D} = \frac{C(1 - b/100) + M(1 + h/100)}{C(1 - b/100) + M(1 + h/100) - (C + M)(1 - D/P)}$$

or

$$\frac{P(1 - k/100)}{D} = \frac{C(1 - b/100) + M(1 + h/100)}{(Mh/100) - (Cb/100) + (C + M)(D/P)}$$

or

$$1 - \frac{k}{100} = \frac{D\{C(1 - b/100) + M(1 + h/100)\}}{P\{(Mh/100) - (Cb/100) + (C + M)(D/P)\}}$$

or

$$\frac{k}{100} = \frac{(PMh/100) - (PCb/100) + (C + M)D - D\{(C + M) + (Mh/100) - (Cb/100)\}}{(PMh/100) + (C + M)D - (PCb/100)}$$

or

$$\frac{k}{100} = \frac{(PMh/100) - (PCb/100) - (MhD/100) + (CbD/100)}{(PMh/100) + (C + M)D - (PCb/100)}$$

or

$$k = \frac{Mh(P - D) - Cb(P - D)}{(PMh/100) + (C + M)D - (PCb/100)}$$

TABLE 1.4

Percentage Reduction in C and P

$b =$	1	3	5	7
$k =$	1.23	0.89	0.54	0.18

or

$$k = \frac{(P - D)(Mh - Cb)}{(PMh/100) + (C + M)D - (PCb/100)}$$

For $h = 10$, Table 1.4 shows various combinations of b and k.

Example 1.4: In the previous example, the % reduction in C and P have been explored. Now consider the I and P. Let:

h = Percentage increase in M
w = Percentage reduction in I
k = Percentage reduction in P

Now:

$$\sqrt{2DI(1 - w/100)\{C + M(1 + h/100)\}\{1 - D/P(1 - k/100)\}}$$
$$= \sqrt{2DI(C + M)(1 - D/P)}$$

or

$$1 - \frac{w}{100} = \frac{(C + M)(1 - D/P)}{\{C + M(1 + h/100)\}\{1 - D/P(1 - k/100)\}}$$

or

$$\frac{w}{100} = 1 - \frac{(C + M)(1 - D/P)}{\{C + M(1 + h/100)\}\{1 - D/P(1 - k/100)\}}$$

or

$$w = \left[1 - \frac{(C + M)(1 - D/P)}{\{C + M(1 + h/100)\}\{1 - D/P(1 - k/100)\}}\right]*100 \quad (1.18)$$

TABLE 1.5

Percentage Reduction in P and I

$k =$	0.2	0.4	0.6	0.8
$w =$	3.68	3.09	2.49	1.88

For $h = 10$, Table 1.5 shows various combinations of k and w.
  If $w = 0$, then from Eq. (1.18):

$$(C + M)(1 - D/P) = \{C + M(1 + h/100)\}\{1 - D/P(1 - k/100)\}$$

or

$$1 - \frac{D}{P(1 - k/100)} = \frac{(C + M)(1 - D/P)}{C + M(1 + h/100)}$$

or

$$\frac{D}{P(1 - k/100)} = \frac{C + M(1 + h/100) - (C + M)(1 - D/P)}{C + M(1 + h/100)}$$

or

$$\frac{D}{P(1 - k/100)} = \frac{C + M + (Mh/100) - (C + M) + (C + M)(D/P)}{C + M(1 + h/100)}$$

or

$$\frac{D}{P(1 - k/100)} = \frac{(Mh/100) + (C + M)(D/P)}{C + M(1 + h/100)}$$

or

$$\frac{P(1 - k/100)}{D} = \frac{C + M(1 + h/100)}{(Mh/100) + (C + M)(D/P)}$$

or

$$1 - \frac{k}{100} = \frac{DC + MD(1 + h/100)}{(PMh/100) + (C + M)D}$$

or

$$1 - \frac{k}{100} = \frac{(C + M)D + (MDh/100)}{(PMh/100) + (C + M)D}$$

or

$$\frac{k}{100} = \frac{(PMh/100) - (MDh/100)}{(PMh/100) + (C + M)D}$$

or

$$k = \frac{PMh - MDh}{(PMh/100) + (C + M)D}$$

or

$$k = \frac{Mh(P - D)}{(PMh/100) + (C + M)D} \tag{1.19}$$

For $h = 10$:
  $k = 1.40$
  If $k = 0$, then from Eq. (1.18):

$$w = \left[1 - \frac{(C + M)}{C + M(1 + h/100)}\right] * 100$$

or

$$w = \left[\frac{C + M(1 + h/100) - (C + M)}{C + M(1 + h/100)}\right] * 100$$

or

$$w = \left[\frac{(Mh/100)}{C + M(1 + h/100)}\right] * 100$$

or

$$w = \frac{Mh}{C + M(1 + h/100)} \tag{1.20}$$

For $h = 10$:
   $w = 4.25$
   It can be observed that:
   $w > k$
   This is because (using Eqs. (1.19) and (1.20)):

$$\frac{Mh}{C + M(1 + h/100)} > \frac{Mh(P - D)}{(PMh/100) + (C + M)D}$$

or

$$(P - D)\{C + M(1 + h/100)\} < (PMh/100) + (C + M)D$$

or

$$(P - D)\{C + M + (Mh/100)\} < (PMh/100) + (C + M)D$$

or

$$(C + M)(P - D) + (P - D)(Mh/100) < (PMh/100) + (C + M)D$$

or

$$(PMh/100) - (P - D)(Mh/100) + (C + M)D - (C + M)(P - D) > 0$$

or

$$(DMh/100) + (C + M)(D - P + D) > 0$$

or

$$(DMh/100) + (C + M)(2D - P) > 0$$

For all practical purposes, this is true because the condition is easily satisfied even if P is as high as equivalent to 2-D.

Thus, the percentage reduction in I as well as the percentage reduction in P may be explored in addition to a simultaneous reduction in both I and P, depending on the production environment and the feasibility of an implementation.

The forecasting expenditure is included explicitly in the present discussion and has its own utility in the said environment. However, if the potential benefits are also assessed and incorporated in the analysis, then an enhanced utility of the approach might be observed in practice.

1.4 POTENTIAL BENEFIT

An organization runs its normal operations at certain level of forecasting expenditure. However, the business planner may feel that there is a need to enhance the expenditure on forecasting efforts. In order to justify such a proposed increase in the related cost, it is reasonable to assess the potential benefit. Since a relative benefit is more valuable in industrial or business practice, a fractional increase in the forecasting expenditure can easily be justified. A fractional decrease in the variation of actual demand from the forecast is a potential outcome, and therefore should be captured in the assessment of an overall potential benefit.

As the inventory planning is conventionally done on the basis of total cost, an overall potential benefit thus obtained needs to be deducted from the total cost comprising of the relevant cost components. The modified total cost thus arrived, has been used for furthermore analysis in the present book for cases such as the procurement and production.

An important requirement for the forecasting expenditure has been discussed in this Chapter 1, i.e., the introduction. A planned schedule is explained for procurement as well as a manufacturing scenario. In a first of its kind, the problem is narrated and a stage is set toward an approach for justifying the forecasting expenditure. Such a forecasting expenditure is specifically included for the buying case in order to develop the total cost. For the making case also, such an expenditure is incorporated in order to arrive at the relevant total cost. The potential benefit is mentioned briefly so that it can be included in more details in the subsequent related chapters.

Procurement is an important function in an organization of all kinds such as trading, production, and services. Irrespective of the kind of organization, forecasting efforts are of significance and their inclusion has an immense benefit. Chapter 2 is devoted to the procurement function in an organization. After incorporating the potential benefit, suitable comparison has been made in order to arrive at the related index. An appropriate expression is developed to get the relevant threshold value. Significance of the threshold value in this context is discussed and analyzed. The development in this regard is also illustrated with suitable examples in order to get the managerial insights.

Production is an important activity in an industrial organization of all types. Forecasting efforts are of significance and their inclusion has an immense benefit irrespective of the kind of industrial organization. Chapter 3 is devoted to the production function in an organization. After incorporating the potential benefit, suitable comparison has been made in order to arrive at the relevant index. An appropriate expression is developed to get the related threshold value. Significance of the threshold value in this context is discussed and analyzed. The development in this regard is also illustrated with suitable examples in order to get the managerial insights.

When the resources of various kinds are not in equivalency of the listed tasks, then the priority should be decided. In addition to the conventional applications discussed before, priority planning is also necessary in many cases. This is also concerned with the planned shortages indirectly. Such applications are included in Chapter 4, along with examples and analysis. For instance, if the resources are not enough to disallow shortages, the backlog might be planned also. Such planned shortages can be backlogged. However, the forecasting expenditure needs to be incorporated as its explicit inclusion is currently the focus area in the applications.

In order to summarize the proposed approach finally, certain highlights are provided in Chapter 5. Factors influencing the cost of forecasting should be sufficiently taken into consideration for an assessment of the relevant expenditure. A firm comes across different stages in its journey in order to attain the operational excellence and maturity. The developed index utility has been described along with the forecasting expenditure need and relevance in the context of various stages of a firm. Additionally, practical discussions have been made related to the evolution stages of an organization and also the business use in order to conclude the overall benefit of the proposed methodology.

2 Procurement

Procurement is an important function in an organization of all kinds such as:

   i. Trading
  ii. Production
 iii. Services

Irrespective of the kind of organization, forecasting efforts are of significance and their inclusion has an immense benefit.

2.1 COST ESTIMATE

Conventionally ordering and carrying costs are added in order to arrive at a total cost in the procurement function. Forecasting expenditure has been introduced before and thus can be incorporated, as shown in Figure 2.1, in many situations.
   This way, the total related cost can comprise:

   i. Ordering cost
  ii. Carrying cost
 iii. Forecasting expenditure

The sum of such costs is discussed in the previous chapter and is given as follows:

$$E = \frac{D}{Q}(C + M) + \frac{Q}{2}.I$$

where
   D = Annual demand
   Q = Ordering quantity in each cycle
   C = Ordering cost
   M = Estimated forecasting expenditure per cycle
   I = Annual inventory carrying cost per unit

2.2 MODIFIED COST ESTIMATE

Assume that the variation of actual demand from the prediction currently is 9% with a certain level of expenditure on the forecasting process. Obviously the desired scenario is a perfect match of actual demand with that predicted. Variation on both sides, i.e., positive as well as the negative side, leads to undesirable effects. When actual demand is more than the predicted demand, the organization has to exert

DOI: 10.1201/9781003267256-2

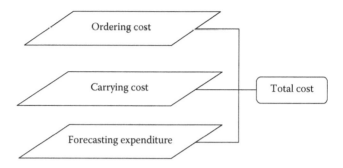

FIGURE 2.1 Cost/expenditure components indicating total cost.

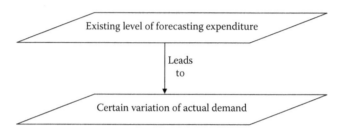

FIGURE 2.2 Existing scenario.

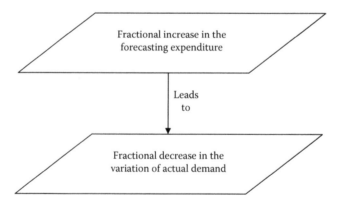

FIGURE 2.3 Proposed scenario.

itself to meet the additional components. In cases where an actual demand is less than predicted, unnecessary material storage and handling might happen. The existing scenario is shown in Figure 2.2.

The management might plan to increase the forecasting expenditure with an anticipated decrease in the variation of actual demand, as shown in Figure 2.3.

For example, if the forecasting expenditure currently is ₹1000 and a proposed plan is to increase this by 10%, then the fractional increase is 0.1 (i.e., the increased component of expenditure is ₹100). It appears reasonable in the management practice to link this increased component with an anticipated benefit. An existing variation of actual demand from the prediction is 9%. It is anticipated that this variation can change to 6% after increasing the expenditure. Thus, the fractional decrease in the variation can be: (9 − 6) = 3%. Since demand is a concerned parameter, this fractional decrease in the variation needs to be linked with the demand. Finally, it should be linked to certain potential benefit per unit improvement (because of a fractional increase in the forecasting expenditure).

In many organizations, current variation of actual demand is easily captured. However, the current level of forecasting expenditure may not be explicit. Thus, the basic case may be considered without any forecasting expenditure inclusion for comparison purposes. Since the fractional increase/decrease in the parameters is captured, the consideration of basic case does not affect the managerial insight.

In order to account for the costs, any overall potential benefit needs to be deducted. Now the modified cost can be expressed as follows:

$$E = \frac{D}{Q}(C + MF) + \frac{Q}{2}. I - SDR \qquad (2.1)$$

where

F = Fractional increase in the forecasting expenditure

R = Potential benefit per unit improvement (because of the increased expenditure on forecasting)

S = Fractional decrease in the variation of actual demand from the forecast

In order to minimize the total cost, differentiate with respect to Q and equate to zero. Thus:

$$\frac{D(C + MF)}{Q^2} = \frac{I}{2}$$

An optimal ordering quantity in each cycle can be expressed as:

$$Q = \sqrt{\frac{2D(C + MF)}{I}}$$

Substituting this optimal value of Q in Eq. (2.1), a minimized total cost can be expressed as:

$$E = \sqrt{2D(C + MF)I} - SDR \qquad (2.2)$$

2.3   RELATED INDEX

As is mentioned previously, a basic case may be considered without any forecasting expenditure inclusion for comparison purposes. It can be shown that the optimal total cost for such a basic case is:

$$E_1 = \sqrt{2DCI} \qquad\qquad (2.3)$$

In order to justify an increased forecasting expenditure,

$$E_1 - E > 0$$

With the substitution of Eqs. (2.2) and (2.3):

$$\sqrt{2DCI} - \sqrt{2D(C + MF)I} + SDR > 0$$

or

$$SDR > \sqrt{2D(C + MF)I} - \sqrt{2DCI}$$

or

$$R > \frac{\sqrt{2DI}}{SD}\left[\sqrt{(C + MF)} - \sqrt{C}\right] \qquad\qquad (2.4)$$

Thus any potential benefit per unit improvement (because of the increased expenditure on forecasting) should be greater than the R.H.S. of Eq. (2.4).

Example 2.1:
Let:

D = 600
C = ₹60
I = ₹20
M = ₹1000
F = 0.1
S = 0.03

From Eq. (2.4):

$$R > \frac{\sqrt{2 \times 600 \times 20}}{0.03 \times 600}\left[\sqrt{(60 + 100)} - \sqrt{60}\right]$$

or

$$R > 42.1995$$

Thus, the threshold value of $R = 42.1995$.

An organization should estimate its R value. If it is less than 42.1995, the proposed increased expenditure on forecasting is not justified. If it is higher than the threshold value, then the proposed increase is justified in a forecasting expenditure. Alternatively, from Eq. (2.4), a related index can be obtained as:

$$\frac{\sqrt{2DI}}{SDR}\left[\sqrt{(C + MF)} - \sqrt{C}\right]$$

For $R = 45$, such an index can be obtained as 0.938. And for $R = 50$, such an index is 0.844. Thus, the lower value of index is preferred relatively in the present discussion. Howeve,r the threshold value of R has a fundamental significance.

A comparison is made here from the basic case. But an organization might consider the significance of forecasting expenditure and might take this into account for procurement lot size and the related cost. In such a case, Eq. (2.1) can be transformed as:

$$E = \frac{D}{Q}(C + M + MF) + \frac{Q}{2}. I - SDR \qquad (2.5)$$

In order to minimize the total cost, differentiate with respect to Q and equate to zero. Thus:

$$\frac{D[C + M(1 + F)]}{Q^2} = \frac{I}{2}$$

An optimal ordering quantity in each cycle can be expressed as:

$$Q = \sqrt{\frac{2D[C + M(1 + F)]}{I}}$$

Substituting this optimal value of Q in Eq. (2.5), a minimized total cost can be expressed as:

$$E = \sqrt{2D[C + M(1 + F)]I} - SDR \qquad (2.6)$$

And Eq. (2.3) can be transformed as follows:

$$E_1 = \sqrt{2D(C + M)I} \qquad (2.7)$$

In order to justify an increased forecasting expenditure:

$$E_1 - E > 0$$

With the substitution of Eqs. (2.6) and (2.7):

$$\sqrt{2D(C + M)I} - \sqrt{2D\{C + M(1 + F)\}I} + SDR > 0$$

or

$$SDR > \sqrt{2D\{C + M(1 + F)\}I} - \sqrt{2D(C + M)I}$$

or

$$R > \frac{\sqrt{2DI}}{SD}\left[\sqrt{\{C + M(1 + F)\}} - \sqrt{(C + M)}\right] \qquad (2.8)$$

With the given data: $R > 12.92$.

Thus, the threshold value of R is lower than the previous scenario.

A related index for such a case can also be obtained as:

$$\frac{\sqrt{2DI}}{SDR}\left[\sqrt{\{C + M(1 + F)\}} - \sqrt{(C + M)}\right]$$

The use of the related index would depend on the stage in which the organization is presently. However, the firms may focus more on forecast accuracy only; thus, there is a rare chance to incorporate the forecasting expenditure explicitly in deciding the procurement lot size and a relevant total cost. Therefore, in order to implement or initiate the proposed approach, Eq. (2.4) can be used for the purpose of obtaining the threshold value of R. However, in an exceptional case where no idea is available initially for the M value, "M*F" can be replaced by a single parameter "A", i.e., an additionally proposed forecasting expenditure.

2.4 ANALYSIS

The threshold value of R has fundamental significance and therefore its variation with respect to a variation in M can be analyzed. Referring Eq. (2.4), the threshold value of R can be expressed as:

$$R = \frac{\sqrt{2DI}}{SD}\left[\sqrt{(C + FM)} - \sqrt{C}\right] \qquad (2.9)$$

2.4.1 REDUCTION IN THE FORECASTING EXPENDITURE

A firm can make efforts to decrease the forecasting expenditure. Now:

h = % reduction in the forecasting expenditure per cycle, M

b = % reduction in the threshold value of R

And, with the use of Eq. (2.9):

$$R\left(1 - \frac{b}{100}\right) = \frac{\sqrt{2DI}}{SD}\left[\sqrt{\{C + FM(1 - h/100)\}} - \sqrt{C}\right]$$

or

$$1 - \frac{b}{100} = \frac{\sqrt{2DI}}{SDR}\left[\sqrt{\{C + FM(1 - h/100)\}} - \sqrt{C}\right]$$

Substituting the value of R from Eq. (2.9):

$$1 - \frac{b}{100} = \frac{\sqrt{\{C + FM(1 - h/100)\}} - \sqrt{C}}{\sqrt{(C + FM)} - \sqrt{C}}$$

or

$$\frac{b}{100} = \frac{\sqrt{(C + FM)} - \sqrt{C} - \sqrt{\{C + FM(1 - h/100)\}} + \sqrt{C}}{\sqrt{(C + FM)} - \sqrt{C}}$$

or

$$b = 100 * \left[\frac{\sqrt{(C + FM)} - \sqrt{\{C + FM(1 - h/100)\}}}{\sqrt{(C + FM)} - \sqrt{C}}\right] \qquad (2.10)$$

With the use of given data from Example 2.1, Table 2.1 shows the percentage decrease in the threshold value of R with respect to the percentage decrease in M.

2.4.2 Increase in the Forecasting Expenditure

Because of unavoidable reasons, there might be an increase in the forecasting expenditure. Now:

$h = \%$ increase in the forecasting expenditure per cycle, M
$b = \%$ increase in the threshold value of R
And, with the use of Eq. (2.9):

TABLE 2.1
Value of b with Reduction in M

$h =$	2	4	6	8	10
$b =$	1.62	3.24	4.88	6.53	8.19

$$R\left(1 + \frac{b}{100}\right) = \frac{\sqrt{2DI}}{SD}\left[\sqrt{\{C + FM(1 + h/100)\}} - \sqrt{C}\right]$$

or

$$1 + \frac{b}{100} = \frac{\sqrt{2DI}}{SDR}\left[\sqrt{\{C + FM(1 + h/100)\}} - \sqrt{C}\right]$$

Substituting the value of R from Eq. (2.9):

$$1 + \frac{b}{100} = \frac{\sqrt{\{C + FM(1 + h/100)\}} - \sqrt{C}}{\sqrt{(C + FM)} - \sqrt{C}}$$

or

$$\frac{b}{100} = \frac{\sqrt{\{C + FM(1 + h/100)\}} - \sqrt{C} - \sqrt{(C + FM)} + \sqrt{C}}{\sqrt{(C + FM)} - \sqrt{C}}$$

or

$$b = 100 * \left[\frac{\sqrt{\{C + FM(1 + h/100)\}} - \sqrt{(C + FM)}}{\sqrt{(C + FM)} - \sqrt{C}}\right] \qquad (2.11)$$

Table 2.2 shows the percentage increase in the threshold value of R with respect to the percentage increase in M.

Values of b are observed to be lower in comparison with the previous situation. Analytically also, it can be shown considering Eqs. (2.10) and (2.11) as:

$$\sqrt{\{C + FM(1 + h/100)\}} - \sqrt{(C + FM)} < \sqrt{(C + FM)}$$
$$- \sqrt{\{C + FM(1 - h/100)\}}$$

or

$$\sqrt{\{C + FM(1 + h/100)\}} + \sqrt{\{C + FM(1 - h/100)\}} < 2\sqrt{(C + FM)}$$

TABLE 2.2

Value of b with Rise in M

$h =$	2	4	6	8	10
$b =$	1.61	3.20	4.79	6.37	7.94

---

or

$$\frac{C + FM\,(1 + h/100) + C + FM\,(1 - h/100) +}{2\sqrt{\{C + FM\,(1 + h/100)\}\{C + FM\,(1 - h/100)\}}} < 4(C + FM)$$

or

$$\frac{2C + FM\,\{1 + (h/100) + 1 - (h/100)\} +}{2\sqrt{\{C + FM\,(1 + h/100)\}\{C + FM\,(1 - h/100)\}}} < 4(C + FM)$$

or

$$2C + 2FM + 2\sqrt{\{C + FM\,(1 + h/100)\}\{C + FM\,(1 - h/100)\}} < 4(C + FM)$$

or

$$(C + FM) + \sqrt{\{C + FM\,(1 + h/100)\}\{C + FM\,(1 - h/100)\}} < 2(C + FM)$$

or

$$\sqrt{\{C + FM\,(1 + h/100)\}\{C + FM\,(1 - h/100)\}} < (C + FM)$$

or

$$\sqrt{C^2 + CFM\,(1 - h/100) + CFM\,(1 + h/100) + F^2M^2\{1 - (h/100)^2\}} < (C + FM)$$

or

$$\sqrt{C^2 + 2CFM + F^2M^2 - (FMh/100)^2} < (C + FM)$$

or

$$\sqrt{(C + FM)^2 - (FMh/100)^2} < (C + FM)$$

or

$$(C + FM)^2 - (FMh/100)^2 < (C + FM)^2$$

And that is true.

Example 2.2: In a case where Eq. (2.8) is relevant, the threshold value of R can be expressed as:

$$R = \frac{\sqrt{2DI}}{SD}\left[\sqrt{\{C + M(1 + F)\}} - \sqrt{(C + M)}\right] \tag{2.12}$$

For a decrease in the forecasting expenditure:

$$R\left(1 - \frac{b}{100}\right) = \frac{\sqrt{2DI}}{SD}\left[\sqrt{\{C + M(1 - h/100)(1 + F)\}} - \sqrt{\{C + M(1 - h/100)\}}\right]$$

or

$$1 - \frac{b}{100} = \frac{\sqrt{2DI}}{SDR}\left[\sqrt{\{C + M(1 - h/100)(1 + F)\}} - \sqrt{\{C + M(1 - h/100)\}}\right]$$

Substituting the value of R from Eq. (2.12):

$$1 - \frac{b}{100} = \frac{\sqrt{\{C + M(1 - h/100)(1 + F)\}} - \sqrt{\{C + M(1 - h/100)\}}}{\sqrt{\{C + M(1 + F)\}} - \sqrt{(C + M)}}$$

or

$$\frac{b}{100} = 1 - \frac{\sqrt{\{C + M(1 - h/100)(1 + F)\}} - \sqrt{\{C + M(1 - h/100)\}}}{\sqrt{\{C + M(1 + F)\}} - \sqrt{(C + M)}}$$

or

$$b = 100 * \left[1 - \frac{\sqrt{\{C + M(1 - h/100)(1 + F)\}} - \sqrt{\{C + M(1 - h/100)\}}}{\sqrt{\{C + M(1 + F)\}} - \sqrt{(C + M)}}\right]$$

With the use of given data from Example 2.1, Table 2.3 shows the percentage decrease in the threshold value of R with respect to the percentage decrease in M.

TABLE 2.3

Value of b with Decrease in M

$h =$	1	2	3	4	5
$b =$	0.53	1.06	1.59	2.13	2.67

For an increase in the forecasting expenditure:

$$R\left(1 + \frac{b}{100}\right) = \frac{\sqrt{2DI}}{SD}\left[\sqrt{\{C + M(1 + h/100)(1 + F)\}} - \sqrt{\{C + M(1 + h/100)\}}\right]$$

or

$$1 + \frac{b}{100} = \frac{\sqrt{2DI}}{SDR}\left[\sqrt{\{C + M(1 + h/100)(1 + F)\}} - \sqrt{\{C + M(1 + h/100)\}}\right]$$

Substituting the value of R from Eq. (2.12):

$$1 + \frac{b}{100} = \frac{\sqrt{\{C + M(1 + h/100)(1 + F)\}} - \sqrt{\{C + M(1 + h/100)\}}}{\sqrt{\{C + M(1 + F)\}} - \sqrt{(C + M)}}$$

or

$$\frac{b}{100} = \frac{\sqrt{\{C + M(1 + h/100)(1 + F)\}} - \sqrt{\{C + M(1 + h/100)\}}}{\sqrt{\{C + M(1 + F)\}} - \sqrt{(C + M)}} - 1$$

or

$$b = 100 * \left[\frac{\sqrt{\{C + M(1 + h/100)(1 + F)\}} - \sqrt{\{C + M(1 + h/100)\}}}{\sqrt{\{C + M(1 + F)\}} - \sqrt{(C + M)}} - 1\right]$$

With the use of given data from Example 2.1, Table 2.4 shows the percentage increase in the threshold value of R with respect to the percentage increase in M.

Values of b are observed to be lower in comparison with the previous situation.

TABLE 2.4

Value of b with Increase in M

h =	1	2	3	4	5
b =	0.52	1.05	1.57	2.09	2.60

2.5 THRESHOLD VALUE

The threshold value of R is expressed by Eq. (2.9). In addition to the forecasting expenditure, other parameters can also influence this value.

2.5.1 ORDERING COST VARIATION

For a reduction in the ordering cost:
h = % reduction in the ordering cost, C
b = % increase in the threshold value of R
And, with the use of expression (2.9):

$$R\left(1 + \frac{b}{100}\right) = \frac{\sqrt{2DI}}{SD}\left[\sqrt{\{C(1 - h/100) + FM\}} - \sqrt{C(1 - h/100)}\right]$$

or

$$1 + \frac{b}{100} = \frac{\sqrt{2DI}}{SDR}\left[\sqrt{\{C(1 - h/100) + FM\}} - \sqrt{C(1 - h/100)}\right]$$

Substituting the value of R from Eq. (2.9):

$$1 + \frac{b}{100} = \frac{\sqrt{\{C(1 - h/100) + FM\}} - \sqrt{C(1 - h/100)}}{\sqrt{(C + FM)} - \sqrt{C}}$$

or

$$b = 100 * \left[\frac{\sqrt{\{C(1 - h/100) + FM\}} - \sqrt{C(1 - h/100)}}{\sqrt{(C + FM)} - \sqrt{C}} - 1\right]$$

With the use of relevant data:
C = ₹60
M = ₹1000
F = 0.1
Table 2.5 shows the percentage increase in the threshold value of R with respect to the percentage decrease in C.

TABLE 2.5

Value of b with Reduction in C

h =	2	4	6	8	10
b =	0.62	1.25	1.89	2.55	3.22

For a rise in the ordering cost:

h = % increase in the ordering cost, C

b = % reduction in the threshold value of R

And, with the use of Eq. (2.9):

$$R\left(1 - \frac{b}{100}\right) = \frac{\sqrt{2DI}}{SD}\left[\sqrt{\{C(1 + h/100) + FM\}} - \sqrt{C(1 + h/100)}\right]$$

or

$$1 - \frac{b}{100} = \frac{\sqrt{2DI}}{SDR}\left[\sqrt{\{C(1 + h/100) + FM\}} - \sqrt{C(1 + h/100)}\right]$$

Substituting the value of R from Eq. (2.9):

$$1 - \frac{b}{100} = \frac{\sqrt{\{C(1 + h/100) + FM\}} - \sqrt{C(1 + h/100)}}{\sqrt{(C + FM)} - \sqrt{C}}$$

or

$$b = 100 * \left[1 - \frac{\sqrt{\{C(1 + h/100) + FM\}} - \sqrt{C(1 + h/100)}}{\sqrt{(C + FM)} - \sqrt{C}}\right]$$

With the use of given data, Table 2.6 shows the percentage reduction in the threshold value of R with respect to the percentage increase in C.

Values of b are observed to be lower in comparison with the previous situation.

TABLE 2.6

Value of b with Increase in C

h =	2	4	6	8	10
b =	0.61	1.20	1.78	2.36	2.92

2.5.2 CARRYING COST VARIATION

Let:

h = % reduction in the carrying cost, I

b = % reduction in the threshold value of R

And, with the use of Eq. (2.9):

$$R\left(1 - \frac{b}{100}\right) = \frac{\sqrt{2DI\,(1 - h/100)}}{SD}\left[\sqrt{(C + FM)} - \sqrt{C}\right]$$

or

$$1 - \frac{b}{100} = \frac{\sqrt{2DI\,(1 - h/100)}}{SDR}\left[\sqrt{(C + FM)} - \sqrt{C}\right]$$

Substituting the value of R from Eq. (2.9):

$$1 - \frac{b}{100} = \sqrt{1 - (h/100)}$$

or

$$\frac{b}{100} = 1 - \sqrt{1 - (h/100)}$$

or

$$b = 100 * \left\{1 - \sqrt{1 - (h/100)}\right\} \tag{2.13}$$

Table 2.7 shows the percentage decrease in the threshold value of R with respect to the percentage decrease in I.

Now, consider:

h = % increase in the carrying cost, I

b = % increase in the threshold value of R

And, with the use of Eq. (2.9):

TABLE 2.7

Value of b with Reduction in I

$h =$	2	4	6	8	10
$b =$	1.00	2.02	3.05	4.08	5.13

$$R\left(1 + \frac{b}{100}\right) = \frac{\sqrt{2DI\,(1 + h/100)}}{SD}\left[\sqrt{(C + FM)} - \sqrt{C}\right]$$

or

$$1 + \frac{b}{100} = \frac{\sqrt{2DI\,(1 + h/100)}}{SDR}\left[\sqrt{(C + FM)} - \sqrt{C}\right]$$

Substituting the value of R from Eq. (2.9):

$$1 + \frac{b}{100} = \sqrt{1 + (h/100)}$$

or

$$\frac{b}{100} = \sqrt{1 + (h/100)} - 1$$

or

$$b = 100 * \left\{\sqrt{1 + (h/100)} - 1\right\} \qquad (2.14)$$

Table 2.8 shows the percentage increase in the threshold value of R with respect to the percentage increase in I.

Values of b are observed to be lower in comparison with the previous situation. Analytically also, it can be shown considering Eqs. (2.13) and (2.14) as:

$$\sqrt{1 + (h/100)} - 1 < 1 - \sqrt{1 - (h/100)}$$

or

$$\sqrt{1 + (h/100)} + \sqrt{1 - (h/100)} < 2$$

or

$$1 + (h/100) + 1 - (h/100) + 2\sqrt{1 - (h/100)^2} < 4$$

TABLE 2.8

Value of b with Increase in I

$h =$	2	4	6	8	10
$b =$	0.99	1.98	2.96	3.92	4.88

or

$$1 + \sqrt{1 - (h/100)^2} < 2$$

or

$$\sqrt{1 - (h/100)^2} < 1$$

As h is less than 100 for all practical purposes, this is true.

After certain awareness and exposure in the context of forecasting expenditure, there would be a possibility for the organizations to include it in the total cost planning and other parameter evaluations also. Influence on the output parameters can also be analyzed following a general approach.

An optimal ordering quantity in each cycle is expressed by Eq. (1.4) as:

$$Q = \sqrt{\frac{2D(C + M)}{I}}$$

A minimized total cost is expressed by Eq. (1.5) as:

$$E = \sqrt{2DI(C + M)}$$

These expressions can be used in order to know the influence of variation in certain input parameters.

Example 2.3: Consider the inventory carrying cost variation. Let:

h = % increase in the carrying cost, I
Therefore, the increased holding cost:

$$I_1 = \left(1 + \frac{h}{100}\right)I$$

$$\text{Decrease in the order size} = \sqrt{\frac{2D(C + M)}{I}} - \sqrt{\frac{2D(C + M)}{I_1}}$$

$$= \sqrt{\frac{2D(C + M)}{I}} - \sqrt{\frac{2D(C + M)}{\left(1 + \frac{h}{100}\right)I}}$$

$$= \sqrt{\frac{2D(C + M)}{I}}\left[1 - \sqrt{\frac{1}{\left(1 + \frac{h}{100}\right)}}\right]$$

And:

$$\% \text{ decrease in } Q = 100 * \left[1 - \sqrt{\frac{1}{\left(1 + \frac{h}{100}\right)}} \right]$$

Now:

$$
\begin{aligned}
\text{Additional cost incurred} &= \sqrt{2D(C + M)h_1} - \sqrt{2D(C + M)I} \\
&= \sqrt{2D(C + M)} \left[\sqrt{h_1} - \sqrt{I} \right] \\
&= \sqrt{2D(C + M)} \left[\sqrt{\left(1 + \frac{h}{100}\right)I} - \sqrt{I} \right] \\
&= \sqrt{2D(C + M)I} \left[\sqrt{\left(1 + \frac{h}{100}\right)} - 1 \right]
\end{aligned}
$$

And:

$$\% \text{ increase in } E = 100 * \left[\sqrt{\left(1 + \frac{h}{100}\right)} - 1 \right]$$

The obtained generalized results are summarized in Table 2.9.

It can be shown that the variation in cost is higher than that in the order size. This is because:

$$\sqrt{\left(1 + \frac{h}{100}\right)} - 1 > 1 - \sqrt{\frac{1}{\left(1 + \frac{h}{100}\right)}}$$

TABLE 2.9
Results with Reference to % Increase in Carrying Cost

Decrease in the order size	$\sqrt{\frac{2D(C + M)}{I}} \left[1 - \sqrt{\frac{1}{\left(1 + \frac{h}{100}\right)}} \right]$
% Decrease in the order size	$100 * \left[1 - \sqrt{\frac{1}{\left(1 + \frac{h}{100}\right)}} \right]$
Additional related cost	$\sqrt{2D(C + M)I} \left[\sqrt{\left(1 + \frac{h}{100}\right)} - 1 \right]$
% Increase in cost	$100 * \left[\sqrt{\left(1 + \frac{h}{100}\right)} - 1 \right]$

---

or

$$2 < \sqrt{\left(1 + \frac{h}{100}\right)} + \sqrt{\frac{1}{\left(1 + \frac{h}{100}\right)}}$$

or

$$4 < \left[\sqrt{\left(1 + \frac{h}{100}\right)} + \sqrt{\frac{1}{\left(1 + \frac{h}{100}\right)}}\right]^2$$

or

$$4 < 1 + \frac{h}{100} + \frac{1}{1 + (h/100)} + 2$$

or

$$4 < 3 + \frac{h}{100} + \frac{1}{1 + (h/100)}$$

or

$$4 < 3 + \frac{h}{100} + \frac{1}{1 + (h/100)}$$

or

$$4 < 3 + \frac{(h/100) + (h/100)^2 + 1}{1 + (h/100)}$$

or

$$4 < 3 + 1 + \frac{(h/100)^2}{1 + (h/100)}$$

or

$$4 < 4 + \frac{(h/100)^2}{1 + (h/100)}$$

And that is true.

In order to illustrate, consider:

Annual demand, D = 600 units

Ordering cost, C = ₹20

Estimated forecasting expenditure per cycle, M = ₹10

Annual inventory carrying cost per unit, I = ₹10

Now:

$$Q = \sqrt{\frac{2D(C+M)}{I}} = 60 \text{ units}$$

$$E = \sqrt{2D(C+M)I} = ₹600$$

The following increase in inventory carrying cost may be implemented:

% Increase in I	5%	10%	15%	20%	25%	30%
I	10.5	11	11.5	12	12.5	13

Because of an increase in carrying cost, the ordering quantity reduces and the related total cost increases. These effects are represented by Table 2.10.

Now consider the reduction in carrying cost, thus:

h = % decrease in inventory carrying cost

Therefore, the decreased carrying cost is:

$$I_1 = \left(1 - \frac{h}{100}\right)I$$

$$\text{Increase in the order size} = \sqrt{\frac{2D(C+M)}{I_1}} - \sqrt{\frac{2D(C+M)}{I}}$$

$$= \sqrt{\frac{2D(C+M)}{\left(1 - \frac{h}{100}\right)I}} - \sqrt{\frac{2D(C+M)}{I}}$$

$$= \sqrt{\frac{2D(C+M)}{I}}\left[\sqrt{\frac{1}{\left(1 - \frac{h}{100}\right)}} - 1\right]$$

TABLE 2.10

Effects on Parameters with Respect to an Increase in Carrying Cost

% Increase in I	5%	10%	15%	20%	25%	30%
I	10.5	11	11.5	12	12.5	13
Q	59	57	56	55	54	53
% Reduction in Q	2.4%	4.7%	6.7%	8.7%	10.6%	12.3%
E	615	629	643	657	671	684
% Increase in E	2.5%	4.9%	7.2%	9.5%	11.8%	14.0%

And:

$$\% \text{ increase in } Q = 100 * \left[\sqrt{\frac{1}{\left(1 - \frac{h}{100}\right)}} - 1 \right]$$

Now:

$$\begin{aligned} \text{Cost reduction} &= \sqrt{2D(C + M)I} - \sqrt{2D(C + M)I_1} \\ &= \sqrt{2D(C + M)} \left[\sqrt{I} - \sqrt{I_1} \right] \\ &= \sqrt{2D(C + M)} \left[\sqrt{I} - \sqrt{\left(1 - \frac{h}{100}\right)I} \right] \\ &= \sqrt{2D(C + M)I} \left[1 - \sqrt{\left(1 - \frac{h}{100}\right)} \right] \end{aligned}$$

And:

$$\% \text{ decrease in } E = 100 * \left[1 - \sqrt{\left(1 - \frac{h}{100}\right)} \right]$$

These generalized results are also summarized in Table 2.11.

TABLE 2.11

Results with Reference to % Decrease in Carrying Cost

Increase in the order size	$\sqrt{\frac{2D(C+M)}{I}} \left[\sqrt{\frac{1}{\left(1 - \frac{h}{100}\right)}} - 1 \right]$
% Increase in the order size	$100_* \left[\sqrt{\frac{1}{\left(1 - \frac{h}{100}\right)}} - 1 \right]$
Cost reduction	$\sqrt{2D(C + M)I} \left[1 - \sqrt{\left(1 - \frac{h}{100}\right)} \right]$
% Reduction in cost	$100_* \left[1 - \sqrt{\left(1 - \frac{h}{100}\right)} \right]$

It can be shown that the variation in order size is higher than that in the cost. This is because:

$$\sqrt{\frac{1}{\left(1 - \frac{h}{100}\right)}} - 1 > 1 - \sqrt{\left(1 - \frac{h}{100}\right)}$$

or

$$2 < \sqrt{\frac{1}{\left(1 - \frac{h}{100}\right)}} + \sqrt{\left(1 - \frac{h}{100}\right)}$$

or

$$4 < \frac{1}{1 - (h/100)} + 1 - \frac{h}{100} + 2$$

or

$$4 < 3 + \frac{1}{1 - (h/100)} - \frac{h}{100}$$

or

$$4 < 3 + \frac{1 - (h/100)(1 - h/100)}{1 - (h/100)}$$

or

$$4 < 3 + \frac{1 - (h/100) + (h/100)^2}{1 - (h/100)}$$

or

$$4 < 3 + 1 + \frac{(h/100)^2}{1 - (h/100)}$$

or

$$4 < 4 + \frac{(h/100)^2}{1 - (h/100)}$$

And that is true for all practical values of $h < 100$.

Consider the earlier mentioned data. In order to illustrate, the reduction in carrying cost is as follows:

% Reduction in I	5%	10%	15%	20%	25%	30%
I	9.5	9	8.5	8	7.5	7

With the reduction in I, an increased ordering quantity can be procured with an overall reduction in the total cost. The results are shown in Table 2.12.

Example 2.4: Consider the demand variation. Let:

$h = \%$ increase in the demand, D
Therefore, the increased demand:

$$D_1 = \left(1 + \frac{h}{100}\right)D$$

$$\text{Increase in the order size} = \sqrt{\frac{2D_1(C+M)}{I}} - \sqrt{\frac{2D(C+M)}{I}}$$

$$= \sqrt{\frac{2D\left(1 + \frac{h}{100}\right)(C+M)}{I}} - \sqrt{\frac{2D(C+M)}{I}}$$

$$= \sqrt{\frac{2D(C+M)}{I}}\left[\sqrt{\left(1 + \frac{h}{100}\right)} - 1\right]$$

And:

$$\% \text{ increase in } Q = 100 * \left[\sqrt{\left(1 + \frac{h}{100}\right)} - 1\right]$$

TABLE 2.12
Effects on Parameters with Respect to a Reduction in Carrying Cost

% Reduction in I	5%	10%	15%	20%	25%	30%
I	9.5	9	8.5	8	7.5	7
Q	62	63	65	67	69	72
% Increase in Q	2.6%	5.4%	8.5%	11.8%	15.5%	19.5%
E	585	569	553	537	520	502
% Reduction in E	2.5%	5.1%	7.8%	10.6%	13.4%	16.3%

Now:

$$\text{Additional cost incurred} = \sqrt{2D_1(C+M)I} - \sqrt{2D(C+M)I}$$
$$= \sqrt{2(C+M)I}\left[\sqrt{D_1} - \sqrt{D}\right]$$
$$= \sqrt{2(C+M)I}\left[\sqrt{\left(1+\frac{h}{100}\right)D} - \sqrt{D}\right]$$
$$= \sqrt{2D(C+M)I}\left[\sqrt{\left(1+\frac{h}{100}\right)} - 1\right]$$

And:

$$\% \text{ increase in E} = 100*\left[\sqrt{\left(1+\frac{h}{100}\right)} - 1\right]$$

The obtained generalized results are summarized in Table 2.13.

Percent variation in the order size as well as the related cost are similar. In order to illustrate, assume the input parameters of the previous example. Consider if an increase in demand is as follows:

Increase in D	5%	10%	15%	20%	25%	30%
D	630	660	690	720	750	780

Table 2.14 represents the approximate variation in parameters with respect to an increase in demand.

With reference to a reduction in demand, let:

$h = \%$ decrease in demand

TABLE 2.13

Results with Reference to % Increase in Demand

Increase in the order size	$\sqrt{\frac{2D(C+M)}{I}}\left[\sqrt{\left(1+\frac{h}{100}\right)} - 1\right]$
% Increase in the order size	$100*\left[\sqrt{\left(1+\frac{h}{100}\right)} - 1\right]$
Additional related cost	$\sqrt{2D(C+M)I}\left[\sqrt{\left(1+\frac{h}{100}\right)} - 1\right]$
% Increase in cost	$100*\left[\sqrt{\left(1+\frac{h}{100}\right)} - 1\right]$

TABLE 2.14

Effects on Parameters with Respect to an Increase in Demand

Increase in D	5%	10%	15%	20%	25%	30%
D	630	660	690	720	750	780
Q	61	63	64	66	67	68
% Increase in Q	2.5%	4.9%	7.2%	9.5%	11.8%	14.0%
E	615	629	643	657	671	684
% Increase in E	2.5%	4.9%	7.2%	9.5%	11.8%	14.0%

The reduced demand:

$$D_1 = \left(1 - \frac{h}{100}\right)D$$

$$\text{Reduction in the order size} = \sqrt{\frac{2D(C+M)}{I}} - \sqrt{\frac{2D_1(C+M)}{I}}$$

$$= \sqrt{\frac{2D(C+M)}{I}} - \sqrt{\frac{2\left(1-\frac{h}{100}\right)D(C+M)}{I}}$$

$$= \sqrt{\frac{2D(C+M)}{I}}\left[1 - \sqrt{\left(1 - \frac{h}{100}\right)}\right]$$

And:

$$\% \text{ reduction in } Q = 100\left[1 - \sqrt{\left(1 - \frac{h}{100}\right)}\right]$$

Now:

$$\text{Cost reduction} = \sqrt{2D(C+M)I} - \sqrt{2D_1(C+M)I}$$

$$= \sqrt{2(C+M)I}\left[\sqrt{D} - \sqrt{D_1}\right]$$

$$= \sqrt{2(C+M)I}\left[\sqrt{D} - \sqrt{\left(1 - \frac{h}{100}\right)D}\right]$$

$$= \sqrt{2D(C+M)I}\left[1 - \sqrt{\left(1 - \frac{h}{100}\right)}\right]$$

And:

$$\% \text{ reduction in } E = 100\left[1 - \sqrt{\left(1 - \frac{h}{100}\right)}\right]$$

TABLE 2.15

Results with Reference to % Reduction in Demand

Reduction in the order size	$\sqrt{\dfrac{2D(C+M)}{I}}\left[1 - \sqrt{\left(1 - \dfrac{h}{100}\right)}\right]$
% Reduction in the order size	$100\left[1 - \sqrt{\left(1 - \dfrac{h}{100}\right)}\right]$
Cost reduction	$\sqrt{2D(C+M)I}\left[1 - \sqrt{\left(1 - \dfrac{h}{100}\right)}\right]$
% Reduction in cost	$100\left[1 - \sqrt{\left(1 - \dfrac{h}{100}\right)}\right]$

The obtained generalized results are summarized in Table 2.15.

Percent variation in the order size as well as the related cost are similar. In order to illustrate, assume the input parameters of the previous example. Consider if a reduction in demand is as follows:

% Reduction in D	5%	10%	15%	20%	25%	30%
D	570	540	510	480	450	420

Approximate variation in parameters is shown in Table 2.16 with reference to a reduction in demand.

The percentage variation in the output parameters can be observed to be higher in comparison with the previous case. Analytically also, it can be shown as follows:

$$1 - \sqrt{1 - (h/100)} > \sqrt{1 + (h/100)} - 1$$

or

$$2 > \sqrt{1 + (h/100)} + \sqrt{1 - (h/100)}$$

TABLE 2.16

Influence on the Parameters with Respect to a Demand Reduction

% Reduction in D	5%	10%	15%	20%	25%	30%
D	570	540	510	480	450	420
Q	58	57	55	54	52	50
% Reduction in Q	2.5%	5.1%	7.8%	10.6%	13.4%	16.3%
E	585	569	553	537	520	502
% Reduction in E	2.5%	5.1%	7.8%	10.6%	13.4%	16.3%

or

$$4 > 1 + (h/100) + 1 - (h/100) + 2\sqrt{1 - (h/100)^2}$$

or

$$4 > 2 + 2\sqrt{1 - (h/100)^2}$$

or

$$2 > 1 + \sqrt{1 - (h/100)^2}$$

or

$$1 > \sqrt{1 - (h/100)^2}$$

As h is less than 100 for all practical purposes, this is true.

Example 2.5:
Let:

Annual demand, D = 600 units
Ordering cost, C = ₹20
Estimated forecasting expenditure per cycle, M = ₹10
Annual inventory carrying cost per unit, I = ₹10
Now, the total cost can be obtained as follows:
$E = \sqrt{2D(C + M)I} = ₹600$
Now, consider the demand increase by 10%, i.e., equivalent to 660 and the corresponding cost:
E = ₹629.29
With the increased demand also, if an aim is to restore the total cost level, then the desired inventory carrying cost can be evaluated as follows:

$$\sqrt{2 \times 660 \times 30I} = 600$$

or
I=₹9.09
Percent reduction in the carrying cost is approximately 9.09% if it is possible to attain by making a focused effort.

In order to generalize:
h = % increase in demand
b = % reduction in the carrying cost

Now:

$$\sqrt{2D(C+M)I} = \sqrt{2D\left(1 + \frac{h}{100}\right)(C+M)I\left(1 - \frac{b}{100}\right)}$$

or

$$1 = \left(1 + \frac{h}{100}\right)\left(1 - \frac{b}{100}\right)$$

or

$$\left(1 - \frac{b}{100}\right) = \frac{1}{(1 + h/100)}$$

or

$$\frac{b}{100} = 1 - \frac{1}{(1 + h/100)}$$

or

$$\frac{b}{100} = \frac{1 + (h/100) - 1}{(1 + h/100)}$$

or

$$\frac{b}{100} = \frac{(h/100)}{(1 + h/100)}$$

or

$$b = \frac{h}{(1 + h/100)}$$

Table 2.17 shows the variation of b with respect to h.

It can be observed that the values of b are relatively lower than the respective values of h. And with the higher values of h, the b values are less sensitive.

In addition to restoring the total related cost, there might be a need to restore the procurement batch size also.

In the context of restoring the batch size with explicit inclusion of forecasting expenditure, a suitable measure or combination can be adopted if it is feasible. Table 2.18 represents a practical guide for furthermore analysis and application.

TABLE 2.17

Variation of b (Carrying Cost) with Respect to h (Demand)

S. No.	h	$b = \frac{h}{(1 + h/100)}$
1	10	9.09
2	20	16.67
3	30	23.08
4	40	28.57
5	50	33.33

TABLE 2.18

A Practical Guide to Restore the Procurement Batch Size

Change Initiated by the Factor	Remedial Measure for Further Analysis
Change initiated by the factor:	Remedial measure for furthermore analysis:
An increase in demand	Ordering cost decrease/Forecasting expenditure decrease
A decrease in demand	Decrease in inventory carrying cost
Ordering cost reduction	An increase in demand/Carrying cost decrease
Forecasting expenditure decrease	An increase in demand/Carrying cost decrease
Carrying cost decrease	Ordering cost decrease/Forecasting expenditure decrease
An increase in carrying cost	An increase in demand

TABLE 2.19

A Practical Guide to Restore the Total Related Procurement Cost

Change Initiated by the Factor	Remedial Measure for Further Analysis
Change initiated by the factor:	Remedial measure for furthermore analysis:
An increase in carrying cost	Ordering cost decrease/Forecasting expenditure decrease
Carrying cost decrease	An increase in demand
An increase in ordering cost	Carrying cost decrease/Forecasting expenditure decrease
An increase in forecasting expenditure	Ordering cost decrease/Carrying cost decrease
Ordering cost decrease	An increase in demand
An increase in demand	Reduction in forecasting/ordering/holding cost
Forecasting expenditure decrease	A potential increase in demand

In the context of restoring the total related cost with explicit inclusion of forecasting expenditure, a suitable measure or combination can be adopted if it is feasible. Table 2.19 represents a practical guide for further analysis and application.

The influence on the significant parameters has been analyzed in different ways in the procurement function of a business organization.

3 Production

Production is an important activity in an industrial organization of all kind,s such as:

   i. Manufacturing
  ii. Pharmaceutical
 iii. Chemical

Irrespective of the kind of industrial organization, forecasting efforts are of significance and their inclusion has an immense benefit.

3.1 TOTAL COST FORMULATION

Conventionally, facility setup and holding costs are added in order to arrive at a total cost in the production-related function. Forecasting expenditure has been introduced before and thus can be incorporated, as shown in Figure 3.1, in many situations.

   This way, the total related cost can comprise of the following:

   i. Setup cost
  ii. Holding cost
 iii. Forecasting expenditure

The sum of such costs is discussed in the Chapter 1 and is given as follows:

$$E = \frac{QI\,(1 - D/P)}{2} + \frac{D}{Q}(C + M)$$

where
    D = Annual demand
    Q = Batch quantity
    C = Facility setup cost
    M = Forecasting expenditure per cycle
    I = Annual inventory holding cost per unit
    P = Annual production rate

3.2 MODIFIED TOTAL COST

As explained in the previous chapter in detail, the basic case may be considered without any forecasting expenditure inclusion for comparison purposes. Since the fractional increase/decrease in the parameters are captured, the consideration of basic case does not affect the managerial insight.

DOI: 10.1201/9781003267256-3

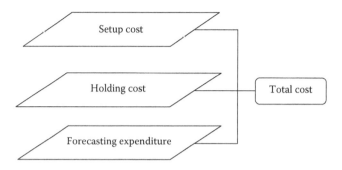

FIGURE 3.1 Components of cost/expenditure indicating total cost.

In order to account for the costs, any overall potential benefit needs to be deducted. Now the modified cost can be expressed as follows:

$$E = \frac{QI(1 - D/P)}{2} + \frac{D}{Q}(C + MF) - SDR \qquad (3.1)$$

where
   F = Fractional increase in the forecasting expenditure
   R = Potential benefit per unit improvement (because of the increased expenditure on forecasting)
   S = Fractional decrease in the variation of actual demand from the forecast
   In order to minimize the total cost, differentiate with respect to Q and equate to zero. Thus:

$$\frac{D(C + MF)}{Q^2} = \frac{I(1 - D/P)}{2}$$

An optimal batch quantity in each production cycle can be expressed as:

$$Q = \sqrt{\frac{2D(C + MF)}{I(1 - D/P)}}$$

Substituting this optimal value of Q in Eq. (3.1), a minimized total cost can be expressed as:

$$E = \sqrt{2D(C + MF)I(1 - D/P)} - SDR \qquad (3.2)$$

3.3   RELEVANT INDEX

As mentioned previously, a basic case may be considered without any forecasting expenditure inclusion for comparison purposes. It can be shown that the optimal total cost for such a basic case is:

$$E_1 = \sqrt{2DCI(1 - D/P)} \qquad (3.3)$$

In order to justify an increased forecasting expenditure:

$$E_1 - E > 0$$

With the substitution of Eqs. (3.2) and (3.3):

$$\sqrt{2DCI(1 - D/P)} - \sqrt{2D(C + MF)I(1 - D/P)} + SDR > 0$$

or

$$SDR > \sqrt{2D(C + MF)I(1 - D/P)} - \sqrt{2DCI(1 - D/P)}$$

or

$$R > \frac{\sqrt{2DI(1 - D/P)}}{SD}\left[\sqrt{(C + MF)} - \sqrt{C}\right] \qquad (3.4)$$

Thus, any potential benefit per unit improvement (because of the increased expenditure on forecasting) should be greater than the R.H.S. of Eq. (3.4).

Example 3.1: Let:

D = 900
P = 1200
C = ₹200
I = ₹50
M = ₹2000
F = 0.1
S = 0.03

From Eq. (3.4):

$$R > \frac{\sqrt{2 \times 900 \times 50 \times (1 - 900/1200)}}{0.03 \times 900}\left[\sqrt{(200 + 200)} - \sqrt{200}\right]$$

or

$$R > 32.54$$

Thus, the threshold value of $R = 32.54$.

An organization should estimate its R value. If it is less than 32.54, the proposed increased expenditure on forecasting is not justified. If it is much higher than the threshold value, the more justified is the proposed increase in a forecasting expenditure. Alternatively, from Eq. (3.4), a relevant index can be obtained as:

$$\frac{\sqrt{2DI(1 - D/P)}}{SDR}\left[\sqrt{(C + MF)} - \sqrt{C}\right]$$

For $R = 40$, such an index can be obtained as 0.81. And for $R = 50$, such an index is 0.65. Thus, the lower value of the index is preferred relatively in the present discussion. However, the threshold value of R has a fundamental significance.

A comparison is made here from the basic case, but an organization might consider the significance of forecasting expenditure and might take this into account for procurement lot size and the related cost. In such a case, Eq. (3.1) can be transformed as:

$$E = \frac{QI(1 - D/P)}{2} + \frac{D}{Q}(C + M + MF) - SDR \qquad (3.5)$$

In order to minimize the total cost, differentiate with respect to Q and equate to zero. Thus:

$$\frac{D[C + M(1 + F)]}{Q^2} = \frac{I(1 - D/P)}{2}$$

An optimal batch quantity in each cycle can be expressed as:

$$Q = \sqrt{\frac{2D[C + M(1 + F)]}{I(1 - D/P)}}$$

Substituting this optimal value of Q in Eq. (3.5), a minimized total cost can be expressed as:

$$E = \sqrt{2D[C + M(1 + F)]I(1 - D/P)} - SDR \qquad (3.6)$$

And Eq. (3.3) can be transformed as follows:

$$E_1 = \sqrt{2D(C + M)I(1 - D/P)} \qquad (3.7)$$

In order to justify an increased forecasting expenditure:

$$E_1 - E > 0$$

With the substitution of Eqs. (3.6) and (3.7):

$$\sqrt{2D(C + M)I(1 - D/P)} - \sqrt{2D\{C + M(1 + F)\}I(1 - D/P)} + SDR > 0$$

or

$$SDR > \sqrt{2D\{C + M(1 + F)\}I(1 - D/P)} - \sqrt{2D(C + M)I(1 - D/P)}$$

or

$$R > \frac{\sqrt{2DI(1 - D/P)}}{SD}\left[\sqrt{\{C + M(1 + F)\}} - \sqrt{(C + M)}\right] \qquad (3.8)$$

With the given data: $R > 11.59$.

Thus, the threshold value of R is lower than the previous scenario.

A related index for such a case can also be obtained as:

$$\frac{\sqrt{2DI(1 - D/P)}}{SDR}\left[\sqrt{\{C + M(1 + F)\}} - \sqrt{(C + M)}\right]$$

The use of the related index would depend on the stage in which a production organization is presently. However, the companies may focus more on forecast accuracy only; thus, there is a rare chance to incorporate the forecasting expenditure explicitly in deciding the production lot size and a related total cost. Therefore, in order to implement or initiate the proposed approach, Eq. (3.4) can be used for the purpose of obtaining the threshold value of R.

3.4   EFFECTS

The threshold value of R has fundamental significance and therefore the effects on its value with respect to a change in M can be observed. Refer Eq. (3.4), where the threshold value of R can be expressed as:

$$R = \frac{\sqrt{2DI(1 - D/P)}}{SD}\left[\sqrt{(C + FM)} - \sqrt{C}\right] \qquad (3.9)$$

3.4.1   DECREASE IN THE FORECASTING EXPENDITURE

A firm can make efforts to decrease the forecasting expenditure. Now:

$h = \%$ reduction in the forecasting expenditure per cycle, M

$b = \%$ reduction in the threshold value of R

And, with the use Eq. (3.9):

$$R\left(1 - \frac{b}{100}\right) = \frac{\sqrt{2DI\,(1 - D/P)}}{SD}\left[\sqrt{\{C + FM\,(1 - h/100)\}} - \sqrt{C}\right]$$

or

$$1 - \frac{b}{100} = \frac{\sqrt{2DI\,(1 - D/P)}}{SDR}\left[\sqrt{\{C + FM\,(1 - h/100)\}} - \sqrt{C}\right]$$

Substituting the value of R from Eq. (3.9):

$$1 - \frac{b}{100} = \frac{\sqrt{\{C + FM\,(1 - h/100)\}} - \sqrt{C}}{\sqrt{(C + FM)} - \sqrt{C}}$$

or

$$\frac{b}{100} = \frac{\sqrt{(C + FM)} - \sqrt{C} - \sqrt{\{C + FM\,(1 - h/100)\}} + \sqrt{C}}{\sqrt{(C + FM)} - \sqrt{C}}$$

or

$$b = 100 * \left[\frac{\sqrt{(C + FM)} - \sqrt{\{C + FM\,(1 - h/100)\}}}{\sqrt{(C + FM)} - \sqrt{C}}\right] \qquad (3.10)$$

Example 3.2: With the use of data from Example 3.1, Table 3.1 shows the percentage reduction in the threshold value of R with respect to the percentage decrease in M.

3.4.2 RAISE IN THE FORECASTING EXPENDITURE

Because of unavoidable reasons, there might be an increase in the forecasting expenditure. Now:

h = % increase in the forecasting expenditure per cycle, M
b = % increase in the threshold value of R

TABLE 3.1
Value of b with Decrease in M

h =	2	4	6	8	10
b =	1.71	3.43	5.16	6.90	8.64

And, with the use Eq. (3.9):

$$R\left(1 + \frac{b}{100}\right) = \frac{\sqrt{2DI(1 - D/P)}}{SD}\left[\sqrt{\{C + FM(1 + h/100)\}} - \sqrt{C}\right]$$

or

$$1 + \frac{b}{100} = \frac{\sqrt{2DI(1 - D/P)}}{SDR}\left[\sqrt{\{C + FM(1 + h/100)\}} - \sqrt{C}\right]$$

Substituting the value of R from Eq. (3.9):

$$1 + \frac{b}{100} = \frac{\sqrt{\{C + FM(1 + h/100)\}} - \sqrt{C}}{\sqrt{(C + FM)} - \sqrt{C}}$$

or

$$\frac{b}{100} = \frac{\sqrt{\{C + FM(1 + h/100)\}} - \sqrt{C} - \sqrt{(C + FM)} + \sqrt{C}}{\sqrt{(C + FM)} - \sqrt{C}}$$

or

$$b = 100 * \left[\frac{\sqrt{\{C + FM(1 + h/100)\}} - \sqrt{(C + FM)}}{\sqrt{(C + FM)} - \sqrt{C}}\right] \quad (3.11)$$

Table 3.2 shows the percentage increase in the threshold value of R with respect to the percentage increase in M.

Values of b are observed to be lower in comparison with the previous situation. Analytically, also, it can be shown considering Eqs. (3.10) and (3.11), as:

$$\sqrt{\{C + FM(1 + h/100)\}} - \sqrt{(C + FM)} < \sqrt{(C + FM)} - \sqrt{\{C + FM(1 - h/100)\}}$$

or

$$\sqrt{\{C + FM(1 + h/100)\}} + \sqrt{\{C + FM(1 - h/100)\}} < 2\sqrt{(C + FM)}$$

TABLE 3.2

Value of b with Increase in M

$h =$	2	4	6	8	10
$b =$	1.70	3.40	5.08	6.76	8.43

or

$$C + FM(1 + h/100) + C + FM(1 - h/100) +$$
$$2\sqrt{\{C + FM(1 + h/100)\}\{C + FM(1 - h/100)\}} < 4(C + FM)$$

or

$$2C + FM\{1 + (h/100) + 1 - (h/100)\} +$$
$$2\sqrt{\{C + FM(1 + h/100)\}\{C + FM(1 - h/100)\}} < 4(C + FM)$$

or

$$2C + 2FM + 2\sqrt{\{C + FM(1 + h/100)\}\{C + FM(1 - h/100)\}} < 4(C + FM)$$

or

$$(C + FM) + \sqrt{\{C + FM(1 + h/100)\}\{C + FM(1 - h/100)\}} < 2(C + FM)$$

or

$$\sqrt{\{C + FM(1 + h/100)\}\{C + FM(1 - h/100)\}} < (C + FM)$$

or

$$\sqrt{C^2 + CFM(1 - h/100) + CFM(1 + h/100) + F^2M^2\{1 - (h/100)^2\}} < (C + FM)$$

or

$$\sqrt{C^2 + 2CFM + F^2M^2 - (FMh/100)^2} < (C + FM)$$

or

$$\sqrt{(C + FM)^2 - (FMh/100)^2} < (C + FM)$$

or

$$(C + FM)^2 - (FMh/100)^2 < (C + FM)^2$$

And that is true.

Example 3.3: In the case where Eq. (3.8) is relevant, the threshold value of R can be expressed as:

$$R = \frac{\sqrt{2DI(1 - D/P)}}{SD}\left[\sqrt{\{C + M(1 + F)\}} - \sqrt{(C + M)}\right] \qquad (3.12)$$

For a decrease in the forecasting expenditure:

$$R\left(1 - \frac{b}{100}\right) = \frac{\sqrt{2DI(1 - D/P)}}{SD}\left[\sqrt{\{C + M(1 - h/100)(1 + F)\}} - \sqrt{\{C + M(1 - h/100)\}}\right]$$

or

$$1 - \frac{b}{100} = \frac{\sqrt{2DI(1 - D/P)}}{SDR}\left[\sqrt{\{C + M(1 - h/100)(1 + F)\}} - \sqrt{\{C + M(1 - h/100)\}}\right]$$

Substituting the value of R from Eq. (3.12):

$$1 - \frac{b}{100} = \frac{\sqrt{\{C + M(1 - h/100)(1 + F)\}} - \sqrt{\{C + M(1 - h/100)\}}}{\sqrt{\{C + M(1 + F)\}} - \sqrt{(C + M)}}$$

or

$$\frac{b}{100} = 1 - \frac{\sqrt{\{C + M(1 - h/100)(1 + F)\}} - \sqrt{\{C + M(1 - h/100)\}}}{\sqrt{\{C + M(1 + F)\}} - \sqrt{(C + M)}}$$

or

$$b = 100 * \left[1 - \frac{\sqrt{\{C + M(1 - h/100)(1 + F)\}} - \sqrt{\{C + M(1 - h/100)\}}}{\sqrt{\{C + M(1 + F)\}} - \sqrt{(C + M)}}\right]$$

With the use of given data from Example 3.1, Table 3.3 shows the percentage decrease in the threshold value of R with respect to the percentage decrease in M.

For an increase in the forecasting expenditure:

$$R\left(1 + \frac{b}{100}\right) = \frac{\sqrt{2DI(1 - D/P)}}{SD}\left[\sqrt{\{C + M(1 + h/100)(1 + F)\}} - \sqrt{\{C + M(1 + h/100)\}}\right]$$

TABLE 3.3

Value of *b* with a Reduction in *M*

$h =$	1	2	3	4	5
$b =$	0.545	1.09	1.64	2.20	2.75

or

$$1 + \frac{b}{100} = \frac{\sqrt{2DI(1 - D/P)}}{SDR}\left[\sqrt{\{C + M(1 + h/100)(1 + F)\}} - \sqrt{\{C + M(1 + h/100)\}}\right]$$

Substituting the value of *R* from Eq. (3.12):

$$1 + \frac{b}{100} = \frac{\sqrt{\{C + M(1 + h/100)(1 + F)\}} - \sqrt{\{C + M(1 + h/100)\}}}{\sqrt{\{C + M(1 + F)\}} - \sqrt{(C + M)}}$$

or

$$\frac{b}{100} = \frac{\sqrt{\{C + M(1 + h/100)(1 + F)\}} - \sqrt{\{C + M(1 + h/100)\}}}{\sqrt{\{C + M(1 + F)\}} - \sqrt{(C + M)}} - 1$$

or

$$b = 100 * \left[\frac{\sqrt{\{C + M(1 + h/100)(1 + F)\}} - \sqrt{\{C + M(1 + h/100)\}}}{\sqrt{\{C + M(1 + F)\}} - \sqrt{(C + M)}} - 1\right]$$

With the use of given data from Example 3.1, Table 3.4 shows the percentage increase in the threshold value of *R* with respect to the percentage increase in *M*. Values of *b* are observed to be lower in comparison with the previous situation.

TABLE 3.4

Value of *b* with a Raise in *M*

$h =$	1	2	3	4	5
$b =$	0.542	1.08	1.62	2.15	2.68

3.5 INFLUENCE ON *R* VALUE

The threshold value of *R* is expressed by Eq. (3.9). In addition to the forecasting expenditure, other parameters can also influence this value.

3.5.1 FACILITY SETUP COST

For a reduction in the setup cost:
h = % reduction in the setup cost, C
b = % increase in the threshold value of R
And, with the use Eq. (3.9):

$$R\left(1 + \frac{b}{100}\right) = \frac{\sqrt{2DI(1 - D/P)}}{SD}\left[\sqrt{\{C(1 - h/100) + FM\}} - \sqrt{C(1 - h/100)}\right]$$

or

$$1 + \frac{b}{100} = \frac{\sqrt{2DI(1 - D/P)}}{SDR}\left[\sqrt{\{C(1 - h/100) + FM\}} - \sqrt{C(1 - h/100)}\right]$$

Substituting the value of *R* from Eq. (3.9):

$$1 + \frac{b}{100} = \frac{\sqrt{\{C(1 - h/100) + FM\}} - \sqrt{C(1 - h/100)}}{\sqrt{(C + FM)} - \sqrt{C}}$$

or

$$b = 100 * \left[\frac{\sqrt{\{C(1 - h/100) + FM\}} - \sqrt{C(1 - h/100)}}{\sqrt{(C + FM)} - \sqrt{C}} - 1\right]$$

With the use of given relevant data as:
C = ₹200
M = ₹2000
F = 0.1
Table 3.5 shows the percentage increase in the threshold value of *R* with respect to the percentage decrease in C.

TABLE 3.5

Value of *b* with Decrease in C

h =	2	4	6	8	10
b =	0.71	1.45	2.19	2.96	3.74

For an increase in the setup cost:
h = % increase in the setup cost, C
b = % reduction in the threshold value of R
And, with the use Eq. (3.9):

$$R\left(1 - \frac{b}{100}\right) = \frac{\sqrt{2DI(1 - D/P)}}{SD}\left[\sqrt{\{C(1 + h/100) + FM\}} - \sqrt{C(1 + h/100)}\right]$$

or

$$1 - \frac{b}{100} = \frac{\sqrt{2DI(1 - D/P)}}{SDR}\left[\sqrt{\{C(1 + h/100) + FM\}} - \sqrt{C(1 + h/100)}\right]$$

Substituting the value of R from Eq. (3.9):

$$1 - \frac{b}{100} = \frac{\sqrt{\{C(1 + h/100) + FM\}} - \sqrt{C(1 + h/100)}}{\sqrt{(C + FM)} - \sqrt{C}}$$

or

$$b = 100 * \left[1 - \frac{\sqrt{\{C(1 + h/100) + FM\}} - \sqrt{C(1 + h/100)}}{\sqrt{(C + FM)} - \sqrt{C}}\right]$$

With the use of given data, Table 3.6 shows the percent reduction in the threshold value of R with respect to the percent increase in C.

Values of b are observed to be lower in comparison with the previous situation.

3.5.2   HOLDING COST

Let:
h = % reduction in the holding cost, I
b = % reduction in the threshold value of R

TABLE 3.6

Value of b with Increase in C

$h =$	2	4	6	8	10
$b =$	0.70	1.38	2.05	2.71	3.35

And, with the use of Eq. (3.9):

$$R\left(1 - \frac{b}{100}\right) = \frac{\sqrt{2DI\,(1 - h/100)(1 - D/P)}}{SD}\left[\sqrt{(C + FM)} - \sqrt{C}\right]$$

or

$$1 - \frac{b}{100} = \frac{\sqrt{2DI\,(1 - h/100)(1 - D/P)}}{SDR}\left[\sqrt{(C + FM)} - \sqrt{C}\right]$$

Substituting the value of R from Eq. (3.9):

$$1 - \frac{b}{100} = \sqrt{1 - (h/100)}$$

or

$$\frac{b}{100} = 1 - \sqrt{1 - (h/100)}$$

or

$$b = 100 * \left\{1 - \sqrt{1 - (h/100)}\right\} \tag{3.13}$$

Table 3.7 shows the percentage reduction in the threshold value of R with respect to the percentage decrease in I.

Now, consider:

$h = \%$ increase in the holding cost, I

$b = \%$ increase in the threshold value of R

And, with the use of Eq. (3.9):

$$R\left(1 + \frac{b}{100}\right) = \frac{\sqrt{2DI\,(1 + h/100)(1 - D/P)}}{SD}\left[\sqrt{(C + FM)} - \sqrt{C}\right]$$

or

$$1 + \frac{b}{100} = \frac{\sqrt{2DI\,(1 + h/100)(1 - D/P)}}{SDR}\left[\sqrt{(C + FM)} - \sqrt{C}\right]$$

TABLE 3.7

Value of b with Decrease in I

$h =$	2	4	6	8	10
$b =$	1.00	2.02	3.05	4.08	5.13

Substituting the value of R from Eq. (3.9):

$$1 + \frac{b}{100} = \sqrt{1 + (h/100)}$$

or

$$\frac{b}{100} = \sqrt{1 + (h/100)} - 1$$

or

$$b = 100 * \left\{ \sqrt{1 + (h/100)} - 1 \right\} \tag{3.14}$$

Table 3.8 shows the percentage increase in the threshold value of R with respect to the percentage raise in I.

Values of b are observed to be lower in comparison with the previous situation. Analytically, also, it can be shown, considering Eqs. (3.13) and (3.14):

$$\sqrt{1 + (h/100)} - 1 < 1 - \sqrt{1 - (h/100)}$$

or

$$\sqrt{1 + (h/100)} + \sqrt{1 - (h/100)} < 2$$

or

$$1 + (h/100) + 1 - (h/100) + 2\sqrt{1 - (h/100)^2} < 4$$

or

$$1 + \sqrt{1 - (h/100)^2} < 2$$

TABLE 3.8

Value of b with Increase in I

$h =$	2	4	6	8	10
$b =$	0.99	1.98	2.96	3.92	4.88

---

or

$$\sqrt{1 - (h/100)^2} < 1$$

As h is less than 100 for all practical purposes, this is true.

After certain awareness and exposure in the context of forecasting expenditure, there would be a possibility for the organizations to include it in the total cost planning and other parameter evaluations also. The influence on the output parameters can also be analyzed following a general approach.

An optimal production batch quantity in each cycle is expressed by Eq. (1.14) as:

$$Q = \sqrt{\frac{2D(C + M)}{I(1 - D/P)}}$$

A minimized total cost is expressed by Eq. (1.15) as:

$$E = \sqrt{2DI(C + M)(1 - D/P)}$$

These expressions can be used in order to know the influence of variation in certain input parameters. For example, consider the inventory carrying cost variation. Let:

h = % increase in the carrying cost, I

Therefore, the increased holding cost:

$$I_1 = \left(1 + \frac{h}{100}\right)I$$

$$\text{Reduction in production batch size} = \sqrt{\frac{2D(C + M)}{(1 - D/P)I}} - \sqrt{\frac{2D(C + M)}{(1 - D/P)I_1}}$$

$$= \sqrt{\frac{2D(C + M)}{(1 - D/P)I}}\left[1 - \sqrt{\frac{I}{I_1}}\right]$$

$$= \sqrt{\frac{2D(C + M)}{(1 - D/P)I}}\left[1 - \sqrt{\frac{1}{(1 + h/100)}}\right]$$

And:

$$\% \text{ reduction in } Q = 100 * \left[1 - \sqrt{\frac{1}{(1 + h/100)}}\right]$$

Now:

$$\text{Increase in total cost} = \sqrt{2D(C + M)I_1(1 - D/P)} - \sqrt{2D(C + M)I(1 - D/P)}$$
$$= \sqrt{2D(C + M)I(1 - D/P)}\left[\sqrt{\frac{I_1}{I}} - 1\right]$$
$$= \sqrt{2D(C + M)I(1 - D/P)}\left[\sqrt{(1 + h/100)} - 1\right]$$

And:

$$\% \text{ increase in E} = 100 * \left\{\sqrt{(1 + h/100)} - 1\right\}$$

The obtained generalized results are summarized in Table 3.9.

It can be shown that the % variation in total cost is higher than that in the batch size as follows:

$$\sqrt{(1 + h/100)} - 1 > 1 - \sqrt{\frac{1}{(1 + h/100)}}$$

or

$$\sqrt{(1 + h/100)} + \sqrt{\frac{1}{(1 + h/100)}} > 2$$

or

$$(1 + h/100) + 1 > 2\sqrt{(1 + h/100)}$$

or

$$2 + h/100 > 2\sqrt{(1 + h/100)}$$

TABLE 3.9

Results with Reference to % Increase in the Carrying Cost

Reduction in the production batch size	$\sqrt{\frac{2D(C+M)}{(1-D/P)I}}\left[1 - \sqrt{\frac{1}{(1+h/100)}}\right]$
% Reduction in the production batch size	$100 * \left[1 - \sqrt{\frac{1}{(1+h/100)}}\right]$
Increase in total cost	$\sqrt{2D(C + M)I(1 - D/P)}\left[\sqrt{(1 + h/100)} - 1\right]$
% Icrease in total cost	$100 * \left\{\sqrt{(1 + h/100)} - 1\right\}$

---

or

$$4 + (h/100)^2 + (h/25) > 4(1 + h/100)$$

or

$$4 + (h/100)^2 + (h/25) > 4 + (h/25)$$

or

$$(h/100)^2 > 0$$

And this is true.

Now, let:

h = % decrease in annual inventory holding cost per unit (I)

And:

$$I_1 = \left(1 - \frac{h}{100}\right)I$$

Increase in production batch size $= \sqrt{\dfrac{2D(C+M)}{(1-D/P)I_1}} - \sqrt{\dfrac{2D(C+M)}{(1-D/P)I}}$

$$= \sqrt{\frac{2D(C+M)}{(1-D/P)I}}\left[\sqrt{\frac{I}{I_1}} - 1\right]$$

$$= \sqrt{\frac{2D(C+M)}{(1-D/P)I}}\left[\sqrt{\frac{1}{(1-h/100)}} - 1\right]$$

And:

$$\% \text{ increase in Q} = 100 * \left\{\sqrt{\frac{1}{(1-h/100)}} - 1\right\}$$

Now:

Reduction in total cost $= \sqrt{2D(C+M)I(1-D/P)} - \sqrt{2D(C+M)I_1(1-D/P)}$

$$= \sqrt{2D(C+M)I(1-D/P)}\left[1 - \sqrt{\frac{I_1}{I}}\right]$$

$$= \sqrt{2D(C+M)I(1-D/P)}\left[1 - \sqrt{(1-h/100)}\right]$$

And:

$$\% \text{ reduction in E} = 100 * \left\{1 - \sqrt{(1-h/100)}\right\}$$

TABLE 3.10

Results with Reference to % Decrease in the Carrying Cost

Increase in the production batch size	$\sqrt{\dfrac{2D(C+M)}{(1-D/P)I}}\left[\sqrt{\dfrac{1}{(1-h/100)}}-1\right]$
% Increase in the production batch size	$100*\left\{\sqrt{\dfrac{1}{(1-h/100)}}-1\right\}$
Reduction in total cost	$\sqrt{2D(C+M)I(1-D/P)}\left[1-\sqrt{(1-h/100)}\right]$
% Reduction in total cost	$100*\left\{1-\sqrt{(1-h/100)}\right\}$

The obtained generalized results are also summarized in Table 3.10.

It can be shown that the % variation in batch size is more than that in the total cost as follows:

$$\sqrt{\frac{1}{(1-h/100)}}-1 > 1-\sqrt{(1-h/100)}$$

or

$$\sqrt{(1-h/100)}+\sqrt{\frac{1}{(1-h/100)}} > 2$$

or

$$(1-h/100)+1 > 2\sqrt{(1-h/100)}$$

or

$$2-h/100 > 2\sqrt{(1-h/100)}$$

or

$$4+(h/100)^2-(h/25) > 4(1-h/100)$$

or

$$4+(h/100)^2-(h/25) > 4-(h/25)$$

or

$$(h/100)^2 > 0$$

And this is true.

Now consider another important parameter, i.e., the annual production rate (P), and its variation. Let:

h = % increase in production rate

$$P_1 = P\left(1 + \frac{h}{100}\right)$$

(a) Reduction in production batch size:

$$\sqrt{\frac{2D(C + M)}{(1 - D/P)I}} - \sqrt{\frac{2D(C + M)}{(1 - D/P_1)I}}$$

$$= \sqrt{\frac{2D(C + M)}{(1 - D/P)I}}\left[1 - \sqrt{\frac{(1 - D/P)}{(1 - D/P_1)}}\right]$$

$$= \sqrt{\frac{2D(C + M)}{(1 - D/P)I}}\left[1 - \sqrt{\frac{(1 - D/P)}{1 - \{D/P(1 + h/100)\}}}\right]$$

And:

$$\% \text{ reduction in } Q = \left[1 - \sqrt{\frac{(1 - D/P)}{1 - \{D/P(1 + h/100)\}}}\right] * 100$$

(b) Increase in total cost:

$$\sqrt{2D(C + M)I(1 - D/P_1)} - \sqrt{2D(C + M)I(1 - D/P)}$$

$$= \sqrt{2D(C + M)I(1 - D/P)}\left[\sqrt{\frac{1 - D/P_1}{(1 - D/P)}} - 1\right]$$

$$= \sqrt{2D(C + M)I(1 - D/P)}\left[\sqrt{\frac{1 - \{D/P(1 + h/100)\}}{(1 - D/P)}} - 1\right]$$

And:

$$\% \text{ increase in } E = \left[\sqrt{\frac{1 - \{D/P(1 + h/100)\}}{(1 - D/P)}} - 1\right] * 100$$

TABLE 3.11

Results with Respect to an Increase in the Production Rate

Reduction in the production batch size	$\sqrt{\dfrac{2D(C+M)}{(1-D/P)I}}\left[1-\sqrt{\dfrac{(1-D/P)}{1-\{D/P(1+h/100)\}}}\right]$
% Reduction in the production batch size	$\left[1-\sqrt{\dfrac{(1-D/P)}{1-\{D/P(1+h/100)\}}}\right]*100$
Increase in the total cost	$\sqrt{2D(C+M)I(1-D/P)}\left[\sqrt{\dfrac{1-\{D/P(1+h/100)\}}{(1-D/P)}}-1\right]$
% Increase in the total cost	$\left[\sqrt{\dfrac{1-\{D/P(1+h/100)\}}{(1-D/P)}}-1\right]*100$

Tables 3.11 summarizes these results.
    Now, let:
    h = % decrease in production rate

$$P_1 = P\left(1 - \frac{h}{100}\right)$$

(a) Increase in production batch size:

$$\sqrt{\frac{2D(C+M)}{(1-D/P_1)I}} - \sqrt{\frac{2D(C+M)}{(1-D/P)I}}$$

$$= \sqrt{\frac{2D(C+M)}{(1-D/P)I}}\left[\sqrt{\frac{(1-D/P)}{(1-D/P_1)}}-1\right]$$

$$= \sqrt{\frac{2D(C+M)}{(1-D/P)I}}\left[\sqrt{\frac{(1-D/P)}{1-\{D/P(1-h/100)\}}}-1\right]$$

And:

$$\% \text{ increase in } Q = \left[\sqrt{\frac{(1-D/P)}{1-\{D/P(1-h/100)\}}}-1\right]*100$$

(b) Reduction in total cost:

$$\sqrt{2D(C+M)I(1-D/P)} - \sqrt{2D(C+M)I(1-D/P_1)}$$

$$= \sqrt{2D(C + M)I(1 - D/P)}\left[1 - \sqrt{\frac{1 - D/P_1}{(1 - D/P)}}\right]$$

$$= \sqrt{2D(C + M)I(1 - D/P)}\left[1 - \sqrt{\frac{1 - \{D/P(1 - h/100)\}}{(1 - D/P)}}\right]$$

And:

$$\% \text{ reduction in } E = \left[1 - \sqrt{\frac{1 - \{D/P(1 - h/100)\}}{(1 - D/P)}}\right] * 100$$

Tables 3.12 summarizes these results.

When a production rate is increased, the total cost also increases. In order to restore the total cost, an effort may be made to reduce the carrying cost. Now, let:

$h = \%$ increase in production rate
$b = \%$ reduction in the carrying cost

$$P_1 = P\left(1 + \frac{h}{100}\right)$$

$$I_1 = I\left(1 - \frac{b}{100}\right)$$

Since the related total cost is given as:

$$E = \sqrt{2DI(C + M)(1 - D/P)},$$

$$\sqrt{2DI_1(C + M)(1 - D/P_1)} = \sqrt{2DI(C + M)(1 - D/P)}$$

TABLE 3.12

Results with Respect to a Decrease in the Production Rate

Increase in the production batch size	$\sqrt{\frac{2D(C+M)}{(1-D/P)I}}\left[\sqrt{\frac{(1-D/P)}{1-\{D/P(1-h/100)\}}} - 1\right]$
% Increase in the production batch size	$\left[\sqrt{\frac{(1-D/P)}{1-\{D/P(1-h/100)\}}} - 1\right] * 100$
Reduction in the total cost	$\sqrt{2D(C+M)I(1-D/P)}\left[1 - \sqrt{\frac{1-\{D/P(1-h/100)\}}{(1-D/P)}}\right]$
% Reduction in the total cost	$\left[1 - \sqrt{\frac{1-\{D/P(1-h/100)\}}{(1-D/P)}}\right] * 100$

---

or

$$I_1(1 - D/P_1) = I(1 - D/P)$$

or

$$\frac{I_1}{I} = \frac{(1 - D/P)}{(1 - D/P_1)}$$

or

$$1 - \frac{b}{100} = \frac{(1 - D/P)}{(1 - D/P_1)}$$

or

$$\frac{b}{100} = 1 - \frac{(1 - D/P)}{(1 - D/P_1)}$$

or

$$\frac{b}{100} = \frac{1 - (D/P_1) - 1 + (D/P)}{(1 - D/P_1)}$$

or

$$\frac{b}{100} = \frac{(D/P) - \{D/P(1 + h/100)\}}{1 - \{D/P(1 + h/100)\}}$$

or

$$\frac{b}{100} = \frac{(1 + h/100)(D/P) - (D/P)}{(1 + h/100) - (D/P)}$$

or

$$\frac{b}{100} = \frac{(h/100)(D/P)}{(h/100) + (1 - D/P)}$$

or

$$b = \frac{h(D/P)}{(h/100) + (1 - D/P)}$$

The relevant parameters are as follows:

$D = 900$

$P = 1200$

Table 3.13 shows the percentage reduction in the carrying cost with respect to the percentage increase in P.

With the carrying cost increase, the total cost also increases. In order to restore the total cost, an effort can be made to reduce the production rate. Now:

h = % increase in carrying cost

b = % reduction in the production rate

$$I_1 = I\left(1 + \frac{h}{100}\right)$$

$$P_1 = P\left(1 - \frac{b}{100}\right)$$

And:

$$\sqrt{2DI_1(C + M)(1 - D/P_1)} = \sqrt{2DI(C + M)(1 - D/P)}$$

or

$$I_1(1 - D/P_1) = I(1 - D/P)$$

or

$$(1 - D/P_1)(1 + h/100) = (1 - D/P)$$

or

$$1 - \frac{D}{P_1} = \frac{(1 - D/P)}{(1 + h/100)}$$

or

$$\frac{D}{P_1} = 1 - \frac{(1 - D/P)}{(1 + h/100)}$$

TABLE 3.13

Value of b Corresponding to h (Increase in P)

$h =$	2	4	6	8	10
$b =$	5.56	10.34	14.52	18.18	21.43

or

$$\frac{D}{P_1} = \frac{1 + (h/100) - 1 + (D/P)}{(1 + h/100)}$$

or

$$\frac{D}{P_1} = \frac{(h/100) + (D/P)}{(1 + h/100)}$$

or

$$\frac{P_1}{D} = \frac{(1 + h/100)}{(h/100) + (D/P)}$$

or

$$P\left(1 - \frac{b}{100}\right) = \frac{D(1 + h/100)}{(h/100) + (D/P)}$$

or

$$1 - \frac{b}{100} = \frac{(D/P)(1 + h/100)}{(h/100) + (D/P)}$$

or

$$\frac{b}{100} = 1 - \frac{(D/P)(1 + h/100)}{(h/100) + (D/P)}$$

or

$$\frac{b}{100} = \frac{(h/100) + (D/P) - (D/P)(1 + h/100)}{(h/100) + (D/P)}$$

or

$$\frac{b}{100} = \frac{(h/100) + (D/P) - (D/P) - (D/P)(h/100)}{(h/100) + (D/P)}$$

or

$$\frac{b}{100} = \frac{(h/100) - (D/P)(h/100)}{(h/100) + (D/P)}$$

or

$$\frac{b}{100} = \frac{(h/100)(1 - D/P)}{(h/100) + (D/P)}$$

or

$$b = \frac{h(1 - D/P)}{(h/100) + (D/P)}$$

The relevant parameters are as follows:
  $D = 900$
  $P = 1200$

Table 3.14 shows the percentage reduction in the production rate with respect to the percentage increase in I.

In the production environment, the total cost may increase because of a certain change in a business/operational factor. It is in the interest of the management to restore the total cost if it is feasible by focusing on a suitable factor as an appropriate response mechanism.

Example 3.4: Let:

D = 600
P = 960
C = ₹30
I = ₹40
M = ₹15

An optimal production batch quantity in each cycle is expressed by Eq. (1.14) as:

$$Q = \sqrt{\frac{2D(C + M)}{I(1 - D/P)}}$$

$$= \sqrt{\frac{2 \times 600 \times 45}{40 \times (1 - 600/960)}}$$

= 60 units

TABLE 3.14

Value of *b* Corresponding to *h* (Increase in *I*)

$h =$	2	4	6	8	10
$b =$	0.65	1.27	1.85	2.41	2.94

A minimized total cost is expressed by Eq. (1.15) as:

$$E = \sqrt{2DI\,(C + M)(1 - D/P)}$$

$$= \sqrt{2 \times 600 \times 40 \times 45(1 - 600/960)}$$

$$= ₹900$$

Now consider an increase in demand as follows:

% Increase in D	5%	10%	15%	20%	25%	30%
D	630	660	690	720	750	780

The variations in parameters with reference to a demand increase are represented by Table 3.15.

For a generalization, let:

$h = \%$ variation in parameter

In the present context, h relates to the % increase in demand; therefore, the increased demand:

$$D_1 = \left(1 + \frac{h}{100}\right)D$$

$$\text{Increase in the production batch size} = \sqrt{\frac{2D_1(C + M)}{(1 - D_1/P)I}} - \sqrt{\frac{2D(C + M)}{(1 - D/P)I}}$$

$$= \sqrt{\frac{2D(C + M)}{(1 - D/P)I}}\left[\sqrt{\frac{D_1(1 - D/P)}{D(1 - D_1/P)}} - 1\right]$$

$$= \sqrt{\frac{2D(C + M)}{(1 - D/P)I}}\left[\sqrt{\frac{(1 + h/100)(1 - D/P)}{1 - (D/P)(1 + h/100)}} - 1\right]$$

TABLE 3.15
Influence of Demand Increase on Parameters

% Increase in D	5%	10%	15%	20%	25%	30%
D	630	660	690	720	750	780
Q	64.22	68.93	74.30	80.50	87.83	96.75
% Increase in Q	7.03%	14.89%	23.83%	34.16%	46.39%	61.25%
E	882.96	861.68	835.84	804.98	768.52	725.60
% Reduction in E	1.89%	4.26%	7.13%	10.56%	14.61%	19.38%

And:

$$\% \text{ increase in } Q = \sqrt{\frac{(1 + h/100)(1 - D/P)}{1 - (D/P)(1 + h/100)}} - 1$$

Now:

$$\text{Reduction in the total cost} = \sqrt{2D(C + M)I(1 - D/P)} - \sqrt{2D_1(C + M)I(1 - D_1/P)}$$

$$= \sqrt{2D(C + M)I(1 - D/P)} \left[1 - \sqrt{\frac{D_1(1 - D_1/P)}{D(1 - D/P)}} \right]$$

$$= \sqrt{2D(C + M)I(1 - D/P)}$$

$$\left[1 - \sqrt{\frac{(1 + h/100)\{1 - (D/P)(1 + h/100)\}}{(1 - D/P)}} \right]$$

And:

$$\% \text{ reduction in } E = 1 - \sqrt{\frac{(1 + h/100)\{1 - (D/P)(1 + h/100)\}}{(1 - D/P)}}$$

Table 3.16 summarizes the obtained results.

Now, consider a decrease in D as follows:

% Reduction in D	5%	10%	15%	20%	25%	30%
D	570	540	510	480	450	420

The variations in parameters with reference to a demand reduction are shown in Table 3.17.

For a generalization, let:

$$D_1 = \left(1 - \frac{h}{100} \right) D$$

TABLE 3.16
Results with Respect to a Demand Increase

Increase in the production batch size	$\sqrt{\frac{2D(C + M)}{(1 - D/P)I}} \left[\sqrt{\frac{(1 + h/100)(1 - D/P)}{1 - (D/P)(1 + h/100)}} - 1 \right]$
% Increase in the production batch size	$\sqrt{\frac{(1 + h/100)(1 - D/P)}{1 - (D/P)(1 + h/100)}} - 1$
Reduction in the total cost	$\sqrt{2D(C + M)I(1 - D/P)} \left[1 - \sqrt{\frac{(1 + h/100)\{1 - (D/P)(1 + h/100)\}}{(1 - D/P)}} \right]$
% Reduction in the total cost	$1 - \sqrt{\frac{(1 + h/100)\{1 - (D/P)(1 + h/100)\}}{(1 - D/P)}}$

TABLE 3.17

Influence of Demand Reduction on Parameters

% Reduction in D	5%	10%	15%	20%	25%	30%
D	570	540	510	480	450	420
Q	56.187	52.699	49.477	46.476	43.656	40.988
% Reduction in Q	6.36%	12.17%	17.54%	22.54%	27.24%	31.69%
E	913.03	922.23	927.70	929.52	927.70	922.23
% Increase in E	1.45%	2.47%	3.08%	3.28%	3.08%	2.47%

Decrease in the production batch size $= \sqrt{\dfrac{2D(C+M)}{(1-D/P)I}} - \sqrt{\dfrac{2D_1(C+M)}{(1-D_1/P)I}}$

$$= \sqrt{\frac{2D(C+M)}{(1-D/P)I}}\left[1 - \sqrt{\frac{D_1(1-D/P)}{D(1-D_1/P)}}\right]$$

$$= \sqrt{\frac{2D(C+M)}{(1-D/P)I}}\left[1 - \sqrt{\frac{(1-h/100)(1-D/P)}{1-(D/P)(1-h/100)}}\right]$$

And:

$$\% \text{ decrease in } Q = 1 - \sqrt{\frac{(1-h/100)(1-D/P)}{1-(D/P)(1-h/100)}}$$

Now:

Increase in the total cost $= \sqrt{2D_1(C+M)I(1-D_1/P)} - \sqrt{2D(C+M)I(1-D/P)}$

$$= \sqrt{2D(C+M)I(1-D/P)}\left[\sqrt{\frac{D_1(1-D_1/P)}{D(1-D/P)}} - 1\right]$$

$$= \sqrt{2D(C+M)I(1-D/P)}$$
$$\left[\sqrt{\frac{(1-h/100)\{1-(D/P)(1-h/100)\}}{(1-D/P)}} - 1\right]$$

And:

$$\% \text{ increase in } E = \sqrt{\frac{(1-h/100)\{1-(D/P)(1-h/100)\}}{(1-D/P)}} - 1$$

Table 3.18 summarizes the obtained results.

TABLE 3.18

Results with Respect to a Demand Reduction

Reduction in the production batch size	$\sqrt{\dfrac{2D(C+M)}{(1-D/P)I}}\left[1 - \sqrt{\dfrac{(1-h/100)(1-D/P)}{1-(D/P)(1-h/100)}}\right]$
% Reduction in the production batch size	$1 - \sqrt{\dfrac{(1-h/100)(1-D/P)}{1-(D/P)(1-h/100)}}$
Increase in the total cost	$\sqrt{2D(C+M)I(1-D/P)}\left[\sqrt{\dfrac{(1-h/100)\{1-(D/P)(1-h/100)\}}{(1-D/P)}} - 1\right]$
% Increase in the total cost	$\sqrt{\dfrac{(1-h/100)\{1-(D/P)(1-h/100)\}}{(1-D/P)}} - 1$

4 Applications

In addition to the conventional applications discussed previously, priority planning is also necessary in many cases. This also concerns the planned shortages indirectly. For instance, if the resources are not enough to disallow shortages, a backlog might be planned also. Such planned shortages can be backlogged. However, the forecasting expenditure needs to be incorporated as its explicit inclusion is currently the focus area in the applications.

4.1 BUYING

In the context of a buying cycle, let:

J = Maximum shortage quantity that is backlogged

K = Annual shortage cost per unit

D = Demand rate per year

Q = Ordering quantity in each cycle

As the shortages occur during the time $\frac{J}{D}$, and there are $\frac{D}{Q}$ cycles in a year, the annual shortage cost can be expressed as:

$$\frac{J}{2} \cdot \frac{J}{D} \cdot \frac{D}{Q} \cdot K$$

$$= \frac{J^2 K}{2Q} \tag{4.1}$$

As the average positive inventory is $\frac{(Q-J)}{2}$, an annual cost related to inventory carrying can be expressed as:

$$\frac{(Q-J)}{2} \cdot \frac{(Q-J)}{D} \cdot \frac{D}{Q} \cdot I$$

$$= \frac{I(Q-J)^2}{2Q} \tag{4.2}$$

where

I = Annual inventory holding cost per unit

In each cycle, the ordering cost and certain forecasting expenditure are incurred. An annual cost concerning such cost components can be expressed as:

DOI: 10.1201/9781003267256-4

$$= \frac{D}{Q}(C + M) \qquad (4.3)$$

where

C = Ordering cost

M = Estimated forecasting expenditure per cycle

A total related cost (E) can be expressed after adding Eqs. (4.1), (4.2), and (4.3) as:

$$E = \frac{D(C + M)}{Q} + \frac{I(Q - J)^2}{2Q} + \frac{J^2 K}{2Q}$$

or

$$E = \frac{D(C + M)}{Q} + \frac{I}{2Q}\{Q^2 - 2QJ + J^2\} + \frac{J^2 K}{2Q}$$

or

$$E = \frac{D(C + M)}{Q} + \frac{IQ}{2} - IJ + \frac{J^2(I + K)}{2Q} \qquad (4.4)$$

$\frac{\partial E}{\partial J} = 0$ shows:

$$\frac{J(I + K)}{Q} = I$$

or

$$J = \frac{IQ}{(I + K)} \qquad (4.5)$$

Substituting in Eq. (4.4):

$$E = \frac{D(C + M)}{Q} + \frac{IQ}{2} - \frac{I^2 Q}{(I + K)} + \frac{I^2 Q}{2(I + K)}$$

or

$$E = \frac{D(C + M)}{Q} + \frac{IQ}{2} - \frac{I^2 Q}{2(I + K)}$$

or

$$E = \frac{D(C + M)}{Q} + \frac{IKQ}{2(I + K)} \qquad (4.6)$$

Differentiating with respect to Q and equating to zero shows:

$$\frac{IK}{2(I + K)} = \frac{D(C + M)}{Q^2}$$

or

$$Q = \sqrt{\frac{2D(C + M)(I + K)}{IK}} \qquad (4.7)$$

Substituting this optimal value of Q in Eq. (4.7), a minimized total cost can be expressed as:

$$E = \sqrt{2D(C + M)IK/(I + K)} \qquad (4.8)$$

4.2 MAKING

In the context of the making cycle, let:

    P = Annual production rate

    J = Maximum shortage quantity that is backlogged

    K = Annual shortage cost per unit

    V = Maximum inventory during the cycle

Shortages occur during the time $\frac{J}{D}$ and $\frac{J}{(P-D)}$. This is because the demand exists at the rate of D however a stock is not available, and afterwards this backlog is replenished at the rate of $(P - D)$ in each cycle. As there are $\frac{D}{Q}$ cycles in a year and the average shortage quantity is $\frac{J}{2}$, the annual shortage cost can be expressed as:

$$\frac{J}{2}\left\{\frac{J}{D} + \frac{J}{(P - D)}\right\} \cdot \frac{D}{Q} \cdot K$$

$$= \frac{J}{2}\left\{\frac{JP}{D(P - D)}\right\} \cdot \frac{D}{Q} \cdot K$$

$$= \frac{J^2 K}{2Q(1 - D/P)} \qquad (4.9)$$

As the batch quantity Q is made in each cycle, and it is consumed at rate D, the time for each cycle is $\frac{Q}{D}$. As the shortages happen during the period $\left\{\frac{J}{D} + \frac{J}{(P-D)}\right\}$, the positive inventory is relevant for the period $\frac{Q}{D} - \left\{\frac{J}{D} + \frac{J}{(P-D)}\right\}$. Since the average inventory is $\frac{V}{2}$ and there are $\frac{D}{Q}$ cycles in a year, the annual inventory holding cost can be expressed as:

$$\frac{V}{2}\left[\frac{Q}{D} - \left\{\frac{J}{D} + \frac{J}{(P-D)}\right\}\right] \cdot \frac{D}{Q} \cdot I$$

$$= \frac{V}{2}\left[\frac{Q}{D} - \frac{JP}{D(P-D)}\right] \cdot \frac{DI}{Q}$$

$$= \frac{VI}{2}\left[1 - \frac{J}{Q(1 - D/P)}\right] \tag{4.10}$$

The quantity Q is made at the rate of P and, during the same time $(V + J)$, is replenished at the rate $(P - D)$; therefore:

$$\frac{V+J}{(P-D)} = \frac{Q}{P}$$

or

$$V = Q(1 - D/P) - J$$

Substituting in Eq. (4.10), the annual holding cost can be expressed as:

$$\frac{\{Q(1 - D/P) - J\}I}{2}\left[1 - \frac{J}{Q(1 - D/P)}\right]$$

$$= \frac{I}{2Q(1 - D/P)}\{Q(1 - D/P) - J\}^2$$

$$= \frac{I}{2Q(1 - D/P)}\{Q^2(1 - D/P)^2 - 2QJ(1 - D/P) + J^2\}$$

$$= \frac{IQ(1 - D/P)}{2} - IJ + \frac{IJ^2}{2Q(1 - D/P)} \tag{4.11}$$

Also, in each cycle, the facility setup cost and certain forecasting expenditures are incurred. An annual cost concerning such cost components can be expressed as:

$$= \frac{D}{Q}(C + M) \qquad (4.12)$$

where

C = Setup cost

M = Estimated forecasting expenditure per cycle

Adding Eqs. (4.9), (4.11), and (4.12), the related total cost can be obtained as:

$$E = \frac{D(C + M)}{Q} + \frac{J^2(I + K)}{2Q(1 - D/P)} + \frac{IQ(1 - D/P)}{2} - IJ \qquad (4.13)$$

$\frac{\partial E}{\partial J} = 0$ shows:

$$\frac{J(I + K)}{Q(1 - D/P)} = I$$

or

$$J = \frac{QI(1 - D/P)}{(I + K)} \qquad (4.14)$$

Substituting in Eq. (4.13):

$$E = \frac{D(C + M)}{Q} + \frac{QI^2(1 - D/P)}{2(I + K)} + \frac{IQ(1 - D/P)}{2} - \frac{I^2Q(1 - D/P)}{(I + K)}$$

or

$$E = \frac{D(C + M)}{Q} + \frac{IQ(1 - D/P)}{2} - \frac{I^2Q(1 - D/P)}{2(I + K)}$$

or

$$E = \frac{D(C + M)}{Q} + \frac{IQ(1 - D/P)}{2}\left\{1 - \frac{I}{(I + K)}\right\}$$

or

$$E = \frac{D(C + M)}{Q} + \frac{IQ(1 - D/P)K}{2(I + K)} \qquad (4.15)$$

Differentiating with respect to Q and equating to zero, shows:

$$\frac{IK(1 - D/P)}{2(I + K)} = \frac{D(C + M)}{Q^2}$$

or

$$Q = \sqrt{\frac{2D(C + M)(I + K)}{IK(1 - D/P)}} \tag{4.16}$$

Substituting this optimal value of Q in Eq. (4.15), a minimized total cost can be expressed as:

$$E = \sqrt{2D(C + M)IK(1 - D/P)/(I + K)} \tag{4.17}$$

4.3   ASSOCIATED INDEX

Following the procedure as explained before, the associated index can be derived for both cases, i.e., for buying and making.

4.3.1   BUYING

Eq. (4.8) can be transformed to include the corresponding benefit for an enhancement in the forecast expenditure. Now:

$$E = \sqrt{2D(C + MF)IK/(I + K)} - SDR \tag{4.18}$$

where
  F = Fractional increase in the forecasting expenditure
  R = Potential benefit per unit improvement (because of the increased expenditure on forecasting)
  S = Fractional decrease in the variation of actual demand from the forecast
  As mentioned previously, a basic case may be considered without any forecasting expenditure inclusion for comparison purposes. It can be shown that the optimal total cost for such a basic case is:

$$E_1 = \sqrt{2DCIK/(I + K)} \tag{4.19}$$

In order to justify an increased forecasting expenditure:

$$E_1 - E > 0$$

With the substitution of Eqs. (4.18) and (4.19):

$$\sqrt{2DCIK/(I + K)} - \sqrt{2D(C + MF)IK/(I + K)} + SDR > 0$$

or

$$SDR > \sqrt{2D(C + MF)IK/(I + K)} - \sqrt{2DCIK/(I + K)}$$

or

$$R > \frac{\sqrt{2DIK/(I + K)}}{SD}\left[\sqrt{(C + MF)} - \sqrt{C}\right] \qquad (4.20)$$

Thus, any potential benefit per unit improvement (because of the increased expenditure on forecasting) should be greater than the R.H.S. of Eq. (4.20).

Example 4.1: Let:

D = 600
C = ₹60
I = ₹20
M = ₹1000
F = 0.1
S = 0.03
K = ₹100

From Eq. (4.20):

$$R > \frac{\sqrt{2 \times 600 \times 20 \times 100/(20 + 100)}}{0.03 \times 600}\left[\sqrt{(60 + 100)} - \sqrt{60}\right]$$

or

$$R > 38.52$$

Thus, the threshold value of $R = 38.52$.

An organization should estimate its R value. If it is less than 38.52, the proposed increased expenditure on forecasting is not justified. It it is much higher than the threshold value, the proposed increase in a forecasting expenditure is justified. Alternatively, from Eq. (4.20), a related index can be obtained as:

$$\frac{\sqrt{2DIK/(I + K)}}{SDR}\left[\sqrt{(C + MF)} - \sqrt{C}\right]$$

Referring to Eq. (4.20), the threshold value of R can be expressed as:

$$R = \frac{\sqrt{2DIK/(I + K)}}{SD}\left[\sqrt{(C + MF)} - \sqrt{C}\right] \qquad (4.21)$$

In order to understand the influence of inventory carrying and shortage cost variation on the threshold value of R, each scenario is analyzed. For a variation in the inventory carrying cost, let:

h = % reduction in the carrying cost, I

b = % reduction in the threshold value of R

And, with the use of Eq. (4.21):

$$R\left(1 - \frac{b}{100}\right) = \frac{\sqrt{2DI(1 - h/100)K/\{I(1 - h/100) + K\}}}{SD}\left[\sqrt{(C + MF)} - \sqrt{C}\right]$$

or

$$1 - \frac{b}{100} = \frac{\sqrt{2DI(1 - h/100)K/\{I(1 - h/100) + K\}}}{SDR}\left[\sqrt{(C + MF)} - \sqrt{C}\right]$$

Substituting the value of R from Eq. (4.21):

$$1 - \frac{b}{100} = \frac{\sqrt{2DI(1 - h/100)K/\{I(1 - h/100) + K\}}}{\sqrt{2DIK/(I + K)}}$$

or

$$1 - \frac{b}{100} = \sqrt{\frac{(1 - h/100)(I + K)}{I(1 - h/100) + K}}$$

or

$$\frac{b}{100} = 1 - \sqrt{\frac{(1 - h/100)(I + K)}{I(1 - h/100) + K}}$$

or

$$b = 100 * \left[1 - \sqrt{\frac{(1 - h/100)(I + K)}{I(1 - h/100) + K}}\right]$$

Table 4.1 shows the percentage decrease in the threshold value of R with respect to the percentage decrease in I.

TABLE 4.1

Value of b with Carrying Cost Reduction

$h =$	2	4	6	8	10
$b =$	0.84	1.69	2.56	3.44	4.33

Now, consider:

$h = \%$ increase in the carrying cost, I

$b = \%$ increase in the threshold value of R

And, with the use of Eq. (4.21):

$$R\left(1 + \frac{b}{100}\right) = \frac{\sqrt{2DI(1 + h/100)K/\{I(1 + h/100) + K\}}}{SD}\left[\sqrt{(C + MF)} - \sqrt{C}\right]$$

or

$$1 + \frac{b}{100} = \frac{\sqrt{2DI(1 + h/100)K/\{I(1 + h/100) + K\}}}{SDR}\left[\sqrt{(C + MF)} - \sqrt{C}\right]$$

Substituting the value of R from Eq. (4.21):

$$1 + \frac{b}{100} = \frac{\sqrt{2DI(1 + h/100)K/\{I(1 + h/100) + K\}}}{\sqrt{2DIK/(I + K)}}$$

or

$$1 + \frac{b}{100} = \sqrt{\frac{(1 + h/100)(I + K)}{I(1 + h/100) + K}}$$

or

$$\frac{b}{100} = \sqrt{\frac{(1 + h/100)(I + K)}{I(1 + h/100) + K}} - 1$$

or

$$b = 100 * \left[\sqrt{\frac{(1 + h/100)(I + K)}{I(1 + h/100) + K}} - 1\right]$$

TABLE 4.2

Value of b with Carrying Cost Increase

$h =$	2	4	6	8	10
$b =$	0.83	1.64	2.44	3.24	4.02

Table 4.2 shows the percentage increase in the threshold value of R with respect to the percentage increase in I.

Values of b are observed to be lower in comparison with the previous situation. For a variation in the shortage cost, let:

$h =$ % reduction in the shortage cost, K

$b =$ % reduction in the threshold value of R

And, with the use of Eq. (4.21):

$$R\left(1 - \frac{b}{100}\right) = \frac{\sqrt{2DIK\,(1 - h/100)/\{I + K\,(1 - h/100)\}}}{SD}\left[\sqrt{(C + MF)} - \sqrt{C}\right]$$

or

$$1 - \frac{b}{100} = \frac{\sqrt{2DIK\,(1 - h/100)/\{I + K\,(1 - h/100)\}}}{SDR}\left[\sqrt{(C + MF)} - \sqrt{C}\right]$$

Substituting the value of R from Eq. (4.21):

$$1 - \frac{b}{100} = \frac{\sqrt{2DIK\,(1 - h/100)/\{I + K\,(1 - h/100)\}}}{\sqrt{2DIK/(I + K)}}$$

or

$$1 - \frac{b}{100} = \sqrt{\frac{(1 - h/100)(I + K)}{I + K\,(1 - h/100)}}$$

or

$$\frac{b}{100} = 1 - \sqrt{\frac{(1 - h/100)(I + K)}{I + K\,(1 - h/100)}}$$

or

$$b = 100 * \left[1 - \sqrt{\frac{(1 - h/100)(I + K)}{I + K\,(1 - h/100)}}\right]$$

TABLE 4.3
Value of b with Shortage Cost Reduction

$h =$	2	4	6	8	10
$b =$	0.17	0.35	0.53	0.72	0.91

Table 4.3 shows the percentage decrease in the threshold value of R with respect to the percentage decrease in K.

Now, consider:

h = % increase in the shortage cost, K

b = % increase in the threshold value of R

And, with the use of Eq. (4.21):

$$R\left(1 + \frac{b}{100}\right) = \frac{\sqrt{2DIK\,(1 + h/100)/\{I + K\,(1 + h/100)\}}}{SD}\left[\sqrt{(C + MF)} - \sqrt{C}\right]$$

or

$$1 + \frac{b}{100} = \frac{\sqrt{2DIK\,(1 + h/100)/\{I + K\,(1 + h/100)\}}}{SDR}\left[\sqrt{(C + MF)} - \sqrt{C}\right]$$

Substituting the value of R from Eq. (4.21):

$$1 + \frac{b}{100} = \frac{\sqrt{2DIK\,(1 + h/100)/\{I + K\,(1 + h/100)\}}}{\sqrt{2DIK/(I + K)}}$$

or

$$1 + \frac{b}{100} = \sqrt{\frac{(1 + h/100)(I + K)}{I + K\,(1 + h/100)}}$$

or

$$\frac{b}{100} = \sqrt{\frac{(1 + h/100)(I + K)}{I + K\,(1 + h/100)}} - 1$$

or

$$b = 100 * \left[\sqrt{\frac{(1 + h/100)(I + K)}{I + K\,(1 + h/100)}} - 1\right]$$

TABLE 4.4

Value of b with Shortage Cost Increase

$h =$	2	4	6	8	10
$b =$	0.16	0.32	0.48	0.62	0.77

Table 4.4 shows the percentage increase in the threshold value of R with respect to the percentage increase in K.

Values of b are observed to be lower in comparison with the previous situation.

Other relevant applications would include an estimation of shortage quantities in each cycle. Consider Example 4.1, the procurement batch quantity can be evaluated from Eq. (4.7) as 276 units, approximately. Substituting in Eq. (4.5), the shortage quantities in each cycle can be obtained as 46 units, approximately.

4.3.2 MAKING

Eq. (4.17) can be transformed to include the corresponding benefit for an enhancement in the forecast expenditure. Now:

$$E = \sqrt{2D(C + MF)IK(1 - D/P)/(I + K)} - SDR \tag{4.22}$$

where

F = Fractional increase in the forecasting expenditure

R = Potential benefit per unit improvement (because of the increased expenditure on forecasting)

S = Fractional decrease in the variation of actual demand from the forecast

As mentioned previously, a basic case may be considered without any forecasting expenditure inclusion for comparison purposes. It can be shown that the optimal total cost for such a basic case is:

$$E_1 = \sqrt{2DCIK(1 - D/P)/(I + K)} \tag{4.23}$$

In order to justify an increased forecasting expenditure:

$$E_1 - E > 0$$

With the substitution of Eqs. (4.22) and (4.23):

$$\sqrt{2DCIK(1 - D/P)/(I + K)} - \sqrt{2D(C + MF)IK(1 - D/P)/(I + K)} + SDR > 0$$

or

$$SDR > \sqrt{2D(C + MF)IK(1 - D/P)/(I + K)} - \sqrt{2DCIK(1 - D/P)/(I + K)}$$

or

$$R > \frac{\sqrt{2DIK\,(1 - D/P)/(I + K)}}{SD}\left[\sqrt{(C + MF)} - \sqrt{C}\right] \qquad (4.24)$$

Thus, any potential benefit per unit improvement (because of the increased expenditure on forecasting) should be greater than the R.H.S. of Eq. (4.24).

Example 4.2: Let:

D = 900
P = 1200
C = ₹200
I = ₹50
M = ₹2000
F = 0.1
S = 0.03
K = ₹400

From Eq. (4.24):

$$R > \frac{\sqrt{2 \times 900 \times 50 \times 400(1 - 900/1200)/(50 + 400)}}{0.03 \times 900}\left[\sqrt{(200 + 200)} - \sqrt{200}\right]$$

or

$$R > 30.68$$

Thus, the threshold value of $R = 30.68$.

An organization should estimate its R value. If it is less than 30.68, the proposed increased expenditure on forecasting is not justified. If it is much higher than the threshold value, the proposed increase in a forecasting expenditure is justified. Alternatively, from Eq. (4.24), a related index can be obtained as:

$$\frac{\sqrt{2DIK\,(1 - D/P)/(I + K)}}{SDR}\left[\sqrt{(C + MF)} - \sqrt{C}\right]$$

Referring to Eq. (4.24), the threshold value of R can be expressed as:

$$R = \frac{\sqrt{2DIK\,(1 - D/P)/(I + K)}}{SD}\left[\sqrt{(C + MF)} - \sqrt{C}\right] \qquad (4.25)$$

In order to understand the influence of inventory carrying and shortage cost variation on the threshold value of R, each scenario is analyzed. For a variation in the inventory carrying cost, let:

h = % reduction in the carrying cost, I
b = % reduction in the threshold value of R
And, with the use of Eq. (4.25):

$$R\left(1 - \frac{b}{100}\right) = \frac{\sqrt{2DI\,(1 - h/100)K\,(1 - D/P)/\{I\,(1 - h/100) + K\}}}{SD}\left[\sqrt{(C + MF)} - \sqrt{C}\right]$$

or

$$1 - \frac{b}{100} = \frac{\sqrt{2DI\,(1 - h/100)K\,(1 - D/P)/\{I\,(1 - h/100) + K\}}}{SDR}\left[\sqrt{(C + MF)} - \sqrt{C}\right]$$

Substituting the value of R from Eq. (4.25):

$$1 - \frac{b}{100} = \frac{\sqrt{2DI\,(1 - h/100)K\,(1 - D/P)/\{I\,(1 - h/100) + K\}}}{\sqrt{2DIK\,(1 - D/P)/(I + K)}}$$

or

$$1 - \frac{b}{100} = \sqrt{\frac{(1 - h/100)(I + K)}{I\,(1 - h/100) + K}}$$

or

$$b = 100 * \left[1 - \sqrt{\frac{(1 - h/100)(I + K)}{I\,(1 - h/100) + K}}\right]$$

Table 4.5 shows the percentage decrease in the threshold value of R with respect to the percentage decrease in I.

TABLE 4.5

Value of b with Carrying Cost Decrease

h =	4	8	12	16
b =	1.80	3.65	5.56	7.52

Now, consider:

h = % increase in the carrying cost, I

b = % increase in the threshold value of R

And, with the use of Eq. (4.25):

$$R\left(1 + \frac{b}{100}\right) = \frac{\sqrt{2DI(1 + h/100)K(1 - D/P)/\{I(1 + h/100) + K\}}}{SD}\left[\sqrt{(C + MF)} - \sqrt{C}\right]$$

or

$$1 + \frac{b}{100} = \frac{\sqrt{2DI(1 + h/100)K(1 - D/P)/\{I(1 + h/100) + K\}}}{SDR}\left[\sqrt{(C + MF)} - \sqrt{C}\right]$$

Substituting the value of R from Eq. (4.25):

$$1 + \frac{b}{100} = \frac{\sqrt{2DI(1 + h/100)K(1 - D/P)/\{I(1 + h/100) + K\}}}{\sqrt{2DIK(1 - D/P)/(I + K)}}$$

or

$$1 + \frac{b}{100} = \sqrt{\frac{(1 + h/100)(I + K)}{I(1 + h/100) + K}}$$

or

$$b = 100 * \left[\sqrt{\frac{(1 + h/100)(I + K)}{I(1 + h/100) + K}} - 1\right]$$

Table 4.6 shows the percentage increase in the threshold value of R with respect to the percentage increase in I.

Values of b are observed to be lower in comparison with the previous situation. For a variation in the shortage cost, let:

h = % reduction in the shortage cost, K

b = % reduction in the threshold value of R

TABLE 4.6

Value of b with Carrying Cost Raise

$h =$	4	8	12	16
$b =$	1.75	3.46	5.13	6.76

And, with the use of Eq. (4.25):

$$R\left(1 - \frac{b}{100}\right) = \frac{\sqrt{2DIK\,(1 - h/100)(1 - D/P)/\{I + K\,(1 - h/100)\}}}{SD}\left[\sqrt{(C + MF)} - \sqrt{C}\right]$$

or

$$1 - \frac{b}{100} = \frac{\sqrt{2DIK\,(1 - h/100)(1 - D/P)/\{I + K\,(1 - h/100)\}}}{SDR}\left[\sqrt{(C + MF)} - \sqrt{C}\right]$$

Substituting the value of R from Eq. (4.25):

$$1 - \frac{b}{100} = \frac{\sqrt{2DIK\,(1 - h/100)(1 - D/P)/\{I + K\,(1 - h/100)\}}}{\sqrt{2DIK\,(1 - D/P)/(I + K)}}$$

or

$$1 - \frac{b}{100} = \sqrt{\frac{(1 - h/100)(I + K)}{I + K\,(1 - h/100)}}$$

or

$$b = 100 * \left[1 - \sqrt{\frac{(1 - h/100)(I + K)}{I + K\,(1 - h/100)}}\right]$$

Table 4.7 shows the percentage decrease in the threshold value of R with respect to the percentage decrease in K.

Now, consider:

$h = \%$ increase in the shortage cost, K

$b = \%$ increase in the threshold value of R

And, with the use of Eq. (4.25):

$$R\left(1 + \frac{b}{100}\right) = \frac{\sqrt{2DIK\,(1 + h/100)(1 - D/P)/\{I + K\,(1 + h/100)\}}}{SD}\left[\sqrt{(C + MF)} - \sqrt{C}\right]$$

TABLE 4.7

Value of b with Shortage Cost Decrease

$h =$	4	8	12	16
$b =$	0.23	0.48	0.75	1.04

---

or

$$1 + \frac{b}{100} = \frac{\sqrt{2DIK\,(1 + h/100)(1 - D/P)/\{I + K\,(1 + h/100)\}}}{SDR}\left[\sqrt{(C + MF)} - \sqrt{C}\right]$$

Substituting the value of R from Eq. (4.25):

$$1 + \frac{b}{100} = \frac{\sqrt{2DIK\,(1 + h/100)(1 - D/P)/\{I + K\,(1 + h/100)\}}}{\sqrt{2DIK\,(1 - D/P)/(I + K)}}$$

or

$$1 + \frac{b}{100} = \sqrt{\frac{(1 + h/100)(I + K)}{I + K\,(1 + h/100)}}$$

or

$$b = 100 * \left[\sqrt{\frac{(1 + h/100)(I + K)}{I + K\,(1 + h/100)}} - 1\right]$$

Table 4.8 shows the percentage increase in the threshold value of R with respect to the percentage increase in K.

Values of b are observed to be lower in comparison with the previous situation.

Other relevant applications would include an estimation of backlogged quantities in each cycle. For the making scenario, consider Example 4.2. The manufacturing lot size can be evaluated from Eq. (4.16) as 597 units. Substituting in Eq. (4.14), the backlogged quantities in each cycle can be evaluated as 17 units, approximately.

4.4 INFLUENCE

Among the applications, the associated index and the related effects have been discussed before. However, after a certain awareness and exposure in the context of forecasting expenditures, there would be a possibility for the organizations to include it in the total cost planning and other parameter evaluations also. The influence on the output parameters can also be analyzed by following a general approach.

TABLE 4.8

Value of b with Shortage Cost Raise

$h =$	4	8	12	16
$b =$	0.21	0.41	0.60	0.78

4.4.1 PURCHASING

An optimal ordering quantity or lot size has been obtained before and given by Eq. (4.7) as follows:

$$Q = \sqrt{\frac{2D(C + M)(K + I)}{IK}}$$

Substituting Eq. (4.7) in Eq. (4.5), the optimal shortage or stock out quantities can be obtained as:

$$J = \sqrt{\frac{2D(C + M)I}{K(K + I)}} \tag{4.26}$$

The total cost has been formulated before and expressed by Eq. (4.8) as follows:

$$E = \sqrt{\frac{2D(C + M)IK}{(K + I)}}$$

In order to know the effects of variation in shortage or stock out cost, and following a general approach, let:

h = % variation in the shortage cost, K

First, consider the increase in stock out cost:

$$K_1 = K\left(1 + \frac{h}{100}\right)$$

A decrease in the lot size with reference to an increased K_1:

$$= \sqrt{\frac{2D(C+M)(K+I)}{IK}} - \sqrt{\frac{2D(C+M)(K_1+I)}{IK_1}}$$

$$= \sqrt{\frac{2D(C+M)(K+I)}{IK}}\left[1 - \sqrt{\frac{K(K_1+I)}{K_1(K+I)}}\right]$$

$$= \sqrt{\frac{2D(C+M)(K+I)}{IK}}\left[1 - \sqrt{\frac{K(1+h/100)+I}{(1+h/100)(K+I)}}\right]$$

And:

$$\% \text{ decrease in lot size} = 1 - \sqrt{\frac{K(1 + h/100) + I}{(1 + h/100)(K + I)}}$$

Decrease in the optimum stock out units $= \sqrt{\dfrac{2D(C+M)I}{K(K+I)}} - \sqrt{\dfrac{2D(C+M)I}{K_1(K_1+I)}}$

$$= \sqrt{\frac{2D(C+M)I}{K(K+I)}}\left[1 - \sqrt{\frac{K(K+I)}{K_1(K_1+I)}}\right]$$

$$= \sqrt{\frac{2D(C+M)I}{K(K+I)}}\left[1 - \sqrt{\frac{(K+I)}{(1+h/100)\{K(1+h/100)+I\}}}\right]$$

And:

% decrease in optimum stock out quantity $= 1 - \sqrt{\dfrac{(K+I)}{(1+h/100)\{K(1+h/100)+I\}}}$

Increase in the total related cost:

$$= \sqrt{\frac{2D(C+M)IK_1}{(K_1+I)}} - \sqrt{\frac{2D(C+M)IK}{(K+I)}}$$

$$= \sqrt{\frac{2D(C+M)IK}{(K+I)}}\left[\sqrt{\frac{K_1(K+I)}{K(K_1+I)}} - 1\right]$$

$$= \sqrt{\frac{2D(C+M)IK}{(K+I)}}\left[\sqrt{\frac{(1+h/100)(K+I)}{K(1+h/100)+I}} - 1\right]$$

And:

% increase in cost $= \sqrt{\dfrac{(1+h/100)(K+I)}{K(1+h/100)+I}} - 1$

The derived results are also summarized in Table 4.9.

TABLE 4.9

Results with Reference to an Increase in Stock Out Cost

Decrease in the lot size	$\sqrt{\dfrac{2D(C+M)(K+I)}{IK}}\left[1 - \sqrt{\dfrac{K(1+h/100)+I}{(1+h/100)(K+I)}}\right]$
% Decrease in the lot size	$1 - \sqrt{\dfrac{K(1+h/100)+I}{(1+h/100)(K+I)}}$
Additional related costs	$\sqrt{\dfrac{2D(C+M)IK}{(K+I)}}\left[\sqrt{\dfrac{(1+h/100)(K+I)}{K(1+h/100)+I}} - 1\right]$
% Increase in costs	$\sqrt{\dfrac{(1+h/100)(K+I)}{K(1+h/100)+I}} - 1$
Decrease in the stock out units	$\sqrt{\dfrac{2D(C+M)I}{K(K+I)}}\left[1 - \sqrt{\dfrac{(K+I)}{(1+h/100)\{K(1+h/100)+I\}}}\right]$
% Decrease in the stock out units	$1 - \sqrt{\dfrac{(K+I)}{(1+h/100)\{K(1+h/100)+I\}}}$

Next, consider the reduction in stock out cost:

$$K_1 = K\left(1 - \frac{h}{100}\right)$$

An increase in the lot size with reference to a reduced K_1:

$$= \sqrt{\frac{2D(C + M)(K_1 + I)}{IK_1}} - \sqrt{\frac{2D(C + M)(K + I)}{IK}}$$

$$= \sqrt{\frac{2D(C + M)(K + I)}{IK}}\left[\sqrt{\frac{K(K_1 + I)}{K_1(K + I)}} - 1\right]$$

$$= \sqrt{\frac{2D(C + M)(K + I)}{IK}}\left[\sqrt{\frac{K(1 - h/100) + I}{(1 - h/100)(K + I)}} - 1\right]$$

And:

$$\% \text{ increase in lot size} = \sqrt{\frac{K(1 - h/100) + I}{(1 - h/100)(K + I)}} - 1$$

$$\text{Increase in the optimum stock out units} = \sqrt{\frac{2D(C + M)I}{K_1(K_1 + I)}} - \sqrt{\frac{2D(C + M)I}{K(K + I)}}$$

$$= \sqrt{\frac{2D(C + M)I}{K(K + I)}}\left[\sqrt{\frac{K(K + I)}{K_1(K_1 + I)}} - 1\right]$$

$$= \sqrt{\frac{2D(C + M)I}{K(K + I)}}\left[\sqrt{\frac{(K + I)}{(1 - h/100)\{K(1 - h/100) + I\}}} - 1\right]$$

And:

$$\% \text{ increase in optimum stock out quantity}$$

$$= \sqrt{\frac{(K + I)}{(1 - h/100)\{K(1 - h/100) + I\}}} - 1$$

Decrease in the total related cost:

$$= \sqrt{\frac{2D(C + M)IK}{(K + I)}} - \sqrt{\frac{2D(C + M)IK_1}{(K_1 + I)}}$$

$$= \sqrt{\frac{2D(C + M)IK}{(K + I)}}\left[1 - \sqrt{\frac{K_1(K + I)}{K(K_1 + I)}}\right]$$

$$= \sqrt{\frac{2D(C + M)IK}{(K + I)}}\left[1 - \sqrt{\frac{(1 - h/100)(K + I)}{K(1 - h/100) + I}}\right]$$

And:

$$\% \text{ decrease in cost} = 1 - \sqrt{\frac{(1 - h/100)(K + I)}{K(1 - h/100) + I}}$$

The derived results are also summarized in Table 4.10.

In addition to a variation in stock out or shortage cost, a change in other parameters may also be studied. However, when output parameters, like procurement lot size, changes, they may disrupt the schedule. In certain cases, there may be a need to restore the lot size and suitable measures can be adopted, if feasible. Table 4.11 represents a practical guide for further analysis and application.

TABLE 4.10
Results with Reference to a Reduced Stock Out Cost

Increase in the lot size	$\sqrt{\frac{2D(C+M)(K+I)}{IK}}\left[\sqrt{\frac{K(1-h/100)+I}{(1-h/100)(K+I)}} - 1\right]$
% Increase in the lot size	$\sqrt{\frac{K(1-h/100)+I}{(1-h/100)(K+I)}} - 1$
Decrease in the related costs	$\sqrt{\frac{2D(C+M)IK}{(K+I)}}\left[1 - \sqrt{\frac{(1-h/100)(K+I)}{K(1-h/100)+I}}\right]$
% Decrease in cost	$1 - \sqrt{\frac{(1-h/100)(K+I)}{K(1-h/100)+I}}$
Increase in the stock out units	$\sqrt{\frac{2D(C+M)I}{K(K+I)}}\left[\sqrt{\frac{(K+I)}{(1-h/100)\{K(1-h/100)+I\}}} - 1\right]$
% Increase in the stock out units	$\sqrt{\frac{(K+I)}{(1-h/100)\{K(1-h/100)+I\}}} - 1$

TABLE 4.11
A Practical Guide to Restore the Procurement Lot Size

Change Initiated by the Factor	Remedial Measure for Further Analysis
Change initiated by the factor:	Remedial measure for furthermore analysis:
Ordering cost decrease	An increase in demand/Holding cost decrease/Shortage cost decrease
Reduction in holding cost	Ordering cost reduction/Forecasting expenditure reduction
Raise in holding cost	An increase in demand/Shortage cost decrease
An increase in demand	Ordering cost reduction/Forecasting expenditure reduction
Demand decrease	Holding cost decrease/Shortage cost decrease
Shortage cost increase	An increase in demand/Holding cost decrease
Shortage cost reduction	Ordering cost reduction/Forecasting expenditure reduction
Forecasting expenditure reduction	Holding cost decrease/Shortage cost decrease/An increase in demand

TABLE 4.12

A Practical Guide to Restore the Total Relevant Cost Concerning Purchase

Change Initiated by the Factor	Remedial Measure for Further Analysis
Change initiated by the factor:	Remedial measure for furthermore analysis:
Ordering cost increase	Forecasting expenditure reduction/Holding cost reduction/Shortage cost reduction
Holding cost increase	Forecasting expenditure reduction/Ordering cost reduction/Shortage cost reduction
Increase in demand	Reduction in ordering cost /Reduction in forecasting expenditure/ Holding cost reduction/Shortage cost reduction
Forecasting expenditure increase	Ordering cost reduction/Holding cost reduction/Shortage cost reduction
Shortage cost increase	Ordering cost reduction/Forecasting expenditure reduction/Holding cost reduction

When total relevant cost changes, there can be an aim to restore the cost. Table 4.12 represents a practical guide for further analysis and application.

4.4.2 PRODUCTION

An optimal production batch quantity or lot size has been obtained before and given by Eq. (4.16) as follows:

$$Q = \sqrt{\frac{2D(C + M)(I + K)}{IK(1 - D/P)}}$$

Substituting Eq. (4.16) in Eq. (4.14), the optimal shortage or stock out quantities can be obtained as:

$$J = \sqrt{\frac{2D(C + M)I(1 - D/P)}{K(K + I)}} \tag{4.27}$$

The total cost has been formulated before and expressed by Eq. (4.17) as follows:

$$E = \sqrt{\frac{2D(C + M)IK(1 - D/P)}{(K + I)}}$$

In order to know the effects of variation in shortage or stock out cost, and following a general approach, let:

$h = \%$ variation in the shortage cost, K

First, consider the increase in stock out cost:

$$K_1 = K\left(1 + \frac{h}{100}\right)$$

A decrease in the lot size with reference to an increased K_1:

$$= \sqrt{\frac{2D(C + M)(K + I)}{IK(1 - D/P)}} - \sqrt{\frac{2D(C + M)(K_1 + I)}{IK_1(1 - D/P)}}$$

$$= \sqrt{\frac{2D(C + M)(K + I)}{IK(1 - D/P)}}\left[1 - \sqrt{\frac{K(K_1 + I)}{K_1(K + I)}}\right]$$

$$= \sqrt{\frac{2D(C + M)(K + I)}{IK(1 - D/P)}}\left[1 - \sqrt{\frac{K(1 + h/100) + I}{(1 + h/100)(K + I)}}\right]$$

And:

$$\% \text{ decrease in lot size} = 1 - \sqrt{\frac{K(1 + h/100) + I}{(1 + h/100)(K + I)}}$$

Decrease in the optimum stock out units

$$= \sqrt{\frac{2D(C + M)I(1 - D/P)}{K(K + I)}} - \sqrt{\frac{2D(C + M)I(1 - D/P)}{K_1(K_1 + I)}}$$

$$= \sqrt{\frac{2D(C + M)I(1 - D/P)}{K(K + I)}}\left[1 - \sqrt{\frac{K(K + I)}{K_1(K_1 + I)}}\right]$$

$$= \sqrt{\frac{2D(C + M)I(1 - D/P)}{K(K + I)}}\left[1 - \sqrt{\frac{(K + I)}{(1 + h/100)\{K(1 + h/100) + I\}}}\right]$$

And:

$$\% \text{ decrease in optimum stock out quantity} = 1$$

$$- \sqrt{\frac{(K + I)}{(1 + h/100)\{K(1 + h/100) + I\}}}$$

Increase in the total related cost:

$$= \sqrt{\frac{2D(C+M)IK_1(1-D/P)}{(K_1+I)}} - \sqrt{\frac{2D(C+M)IK(1-D/P)}{(K+I)}}$$

$$= \sqrt{\frac{2D(C+M)IK(1-D/P)}{(K+I)}} \left[\sqrt{\frac{K_1(K+I)}{K(K_1+I)}} - 1 \right]$$

$$= \sqrt{\frac{2D(C+M)IK(1-D/P)}{(K+I)}} \left[\sqrt{\frac{(1+h/100)(K+I)}{K(1+h/100)+I}} - 1 \right]$$

And:

$$\% \text{ increase in cost} = \sqrt{\frac{(1+h/100)(K+I)}{K(1+h/100)+I}} - 1$$

The derived results are also summarized in Table 4.13.

Next, consider the reduction in stock out cost:

$$K_1 = K\left(1 - \frac{h}{100}\right)$$

An increase in the lot size with reference to a reduced K_1:

$$= \sqrt{\frac{2D(C+M)(K_1+I)}{IK_1(1-D/P)}} - \sqrt{\frac{2D(C+M)(K+I)}{IK(1-D/P)}}$$

$$= \sqrt{\frac{2D(C+M)(K+I)}{IK(1-D/P)}} \left[\sqrt{\frac{K(K_1+I)}{K_1(K+I)}} - 1 \right]$$

$$= \sqrt{\frac{2D(C+M)(K+I)}{IK(1-D/P)}} \left[\sqrt{\frac{K(1-h/100)+I}{(1-h/100)(K+I)}} - 1 \right]$$

TABLE 4.13

Results with Reference to an Increase in Stock Out Cost

Decrease in the lot size	$\sqrt{\frac{2D(C+M)(K+I)}{IK(1-D/P)}} \left[1 - \sqrt{\frac{K(1+h/100)+I}{(1+h/100)(K+I)}} \right]$
% Decrease in the lot size	$1 - \sqrt{\frac{K(1+h/100)+I}{(1+h/100)(K+I)}}$
Additional related costs	$\sqrt{\frac{2D(C+M)IK(1-D/P)}{(K+I)}} \left[\sqrt{\frac{(1+h/100)(K+I)}{K(1+h/100)+I}} - 1 \right]$
% Increase in cost	$\sqrt{\frac{(1+h/100)(K+I)}{K(1+h/100)+I}} - 1$
Decrease in the stock out units	$\sqrt{\frac{2D(C+M)I(1-D/P)}{K(K+I)}} \left[1 - \sqrt{\frac{(K+I)}{(1+h/100)\{K(1+h/100)+I\}}} \right]$
% Decrease in the stock out units	$1 - \sqrt{\frac{(K+I)}{(1+h/100)\{K(1+h/100)+I\}}}$

And:

$$\% \text{ increase in lot size} = \sqrt{\frac{K(1 - h/100) + I}{(1 - h/100)(K + I)}} - 1$$

Increase in the optimum stock out units:

$$= \sqrt{\frac{2D(C + M)I(1 - D/P)}{K_1(K_1 + I)}} - \sqrt{\frac{2D(C + M)I(1 - D/P)}{K(K + I)}}$$

$$= \sqrt{\frac{2D(C + M)I(1 - D/P)}{K(K + I)}} \left[\sqrt{\frac{K(K + I)}{K_1(K_1 + I)}} - 1 \right]$$

$$= \sqrt{\frac{2D(C + M)I(1 - D/P)}{K(K + I)}} \left[\sqrt{\frac{(K + I)}{(1 - h/100)\{K(1 - h/100) + I\}}} - 1 \right]$$

And:

$$\% \text{ increase in optimum stock out quantity} = \sqrt{\frac{(K+I)}{(1-h/100)\{K(1-h/100)+I\}}} - 1$$

Decrease in the total related cost:

$$= \sqrt{\frac{2D(C + M)IK(1 - D/P)}{(K + I)}} - \sqrt{\frac{2D(C + M)IK_1(1 - D/P)}{(K_1 + I)}}$$

$$= \sqrt{\frac{2D(C + M)IK(1 - D/P)}{(K + I)}} \left[1 - \sqrt{\frac{K_1(K + I)}{K(K_1 + I)}} \right]$$

$$= \sqrt{\frac{2D(C + M)IK(1 - D/P)}{(K + I)}} \left[1 - \sqrt{\frac{(1 - h/100)(K + I)}{K(1 - h/100) + I}} \right]$$

And:

$$\% \text{ decrease in cost} = 1 - \sqrt{\frac{(1 - h/100)(K + I)}{K(1 - h/100) + I}}$$

The derived results are also summarized in Table 4.14.

In addition to a variation in stock out or shortage cost, a change in other parameters may also be studied. However, when the output parameters, like production lot size, change, they may disrupt the schedule. In certain cases, there may be a need to restore the lot size and suitable measures can be adopted, if feasible. Table 4.15 represents a practical guide for further analysis and application.

When the total relevant cost changes, there can be an aim to restore the cost. Table 4.16 represents a practical guide for further analysis and application.

TABLE 4.14

Results with Reference to a Reduced Stock Out Cost

Increase in the lot size	$\sqrt{\dfrac{2D(C+M)(K+I)}{IK(1-D/P)}}\left[\sqrt{\dfrac{K(1-h/100)+I}{(1-h/100)(K+I)}}-1\right]$
% Increase in the lot size	$\sqrt{\dfrac{K(1-h/100)+I}{(1-h/100)(K+I)}}-1$
Decrease in the related costs	$\sqrt{\dfrac{2D(C+M)IK(1-D/P)}{(K+I)}}\left[1-\sqrt{\dfrac{(1-h/100)(K+I)}{K(1-h/100)+I}}\right]$
% Decrease in cost	$1-\sqrt{\dfrac{(1-h/100)(K+I)}{K(1-h/100)+I}}$
Increase in the stock out units	$\sqrt{\dfrac{2D(C+M)I(1-D/P)}{K(K+I)}}\left[\sqrt{\dfrac{(K+I)}{(1-h/100)\{K(1-h/100)+I\}}}-1\right]$
% Increase in the stock out units	$\sqrt{\dfrac{(K+I)}{(1-h/100)\{K(1-h/100)+I\}}}-1$

TABLE 4.15

A Practical Guide to Restore the Production Lot Size

Change Initiated by the Factor	Remedial Measure for Further Analysis
Change initiated by the factor:	Remedial measure for furthermore analysis:
An increase in demand	Setup cost decrease/Forecasting expenditure decrease/Production rate increase
Demand reduction	Inventory carrying cost decrease/Shortage cost decrease/Production rate reduction
An increase in forecasting expenditure	Setup cost reduction/Production rate increase
An increase in setup cost	Forecasting expenditure decrease/Production rate increase
Forecasting expenditure decrease	Carrying cost decrease/Shortage cost decrease/Production rate reduction/An increase in demand
Setup cost reduction	Production rate reduction/Carrying cost decrease/Shortage cost decrease/An increase in demand
An increase in carrying cost	An increase in demand/Shortage cost decrease/Production rate reduction
Carrying cost reduction	Setup cost reduction/Forecasting expenditure decrease/Production rate increase
An increase in shortage cost	An increase in demand/Carrying cost decrease/Production rate reduction
Shortage cost decrease	Forecasting expenditure decrease/Setup cost decrease/Production rate increase
An increase in production rate	An increase in demand/Carrying cost decrease/Shortage cost decrease
Production rate reduction	Setup cost decrease/Forecasting expenditure decrease

TABLE 4.16
A Practical Guide to Restore the Total Relevant Cost Concerning Production

Change Initiated by the Factor	Remedial Measure for Further Analysis
Change initiated by the factor:	Remedial measure for furthermore analysis:
An increase in forecasting expenditure	Setup cost decrease/Carrying cost decrease/Shortage cost decrease/ Production rate reduction
An increase in setup cost	Forecasting expenditure decrease/Carrying cost decrease/Shortage cost decrease/Production rate reduction
Carrying cost increase	Forecasting expenditure decrease/Setup cost decrease/Production rate reduction/Shortage cost decrease
An increase in shortage cost	Setup cost reduction/Forecasting expenditure decrease/Production rate reduction/Carrying cost decrease
Production rate increase	Forecasting expenditure/Setup/Shortage/Carrying cost decrease

Applications are generalized in the context of priority planning. If a certain activity is not done because of the priority for another, it might be necessary to estimate the stock out or shortage cost in the purchase as well as production function. Along with its inclusion in the inventory planning with forecasting expenditure, certain generalized results are also obtained.

5 Highlights

In order to summarize the proposed approach finally, certain highlights are provided. Additionally, practical discussions have been made related to the evolution stages of an organization and business use. Before proceeding further, it would be appropriate to give some of the developed key formulae.

Firstly, in the context of procurement, a related total cost can be expressed as:

$$E = \sqrt{2D(C + MF)I} - SDR$$

where

D = Annual demand
C = Ordering cost
M = Estimated forecasting expenditure per cycle
I = Annual inventory carrying cost per unit
F = Fractional increase in the forecasting expenditure
R = Potential benefit per unit improvement (because of the increased expenditure on forecasting)
S = Fractional decrease in the variation of actual demand from the forecast

The expression indicating the threshold value is as follows:

$$R > \frac{\sqrt{2DI}}{SD}\left[\sqrt{(C + MF)} - \sqrt{C}\right]$$

And also the expression for related index is provided as:

$$\frac{\sqrt{2DI}}{SDR}\left[\sqrt{(C + MF)} - \sqrt{C}\right]$$

Next, in the context of making, a total cost can be expressed as:

$$E = \sqrt{2D(C + MF)I(1 - D/P)} - SDR$$

where

C = Facility setup cost
P = Annual production rate

The expression indicating the threshold value is as follows:

$$R > \frac{\sqrt{2DI(1 - D/P)}}{SD}\left[\sqrt{(C + MF)} - \sqrt{C}\right]$$

DOI: 10.1201/9781003267256-5

And also the expression for relevant index is provided as:

$$\frac{\sqrt{2DI\,(1 - D/P)}}{SDR}\left[\sqrt{(C + MF)} - \sqrt{C}\right]$$

In the context of priority planning or stock out inclusion, the expressions are developed for buying as well as making applications, such as:

a. Buying:

A total cost can be expressed as:

$$E = \sqrt{2D\,(C + MF)IK/(I + K)} - SDR$$

where
  C = Ordering cost
  K = Annual shortage cost per unit
The expression indicating the threshold value is as follows:

$$R > \frac{\sqrt{2DIK/(I + K)}}{SD}\left[\sqrt{(C + MF)} - \sqrt{C}\right]$$

Also, a relevant index can be provided as:

$$\frac{\sqrt{2DIK/(I + K)}}{SDR}\left[\sqrt{(C + MF)} - \sqrt{C}\right]$$

(b) Making:
  A total cost can be expressed as:

$$E = \sqrt{2D\,(C + MF)IK\,(1 - D/P)/(I + K)} - SDR$$

where
  C = Facility setup cost
  P = Annual production rate
The expression indicating the threshold value is as follows:

$$R > \frac{\sqrt{2DIK\,(1 - D/P)/(I + K)}}{SD}\left[\sqrt{(C + MF)} - \sqrt{C}\right]$$

Also, a relevant index can be provided as:

$$\frac{\sqrt{2DIK\,(1 - D/P)/(I + K)}}{SDR}\left[\sqrt{(C + MF)} - \sqrt{C}\,\right]$$

5.1   COST OF FORECASTING

Factors influencing the cost of forecasting should be sufficiently taken into consideration for an assessment of the relevant expenditure. Some of the factors are listed in Figure 5.1.

The customer base, in general, refers to the spread of consumers. Sub-criteria related to it can include customer segmentation, their income level, spending habits, and any changes in them, among others. The forecasting expenditure depends on the related efforts, and the close examination of factors and sub-criteria associated with it helps in determining the level of effort. This, in turn, is useful for the arrangement of corresponding resources.

Product type and its nature should also be taken into consideration for the assessment of forecasting efforts and the related cost/expenditure. Broadly, this depends on the type of industry, such as automobile, textile, and pharmaceutical among many others. A product might be light or heavy along with a large number of consumers or a limited number of customers/customer companies. The nature of the product may relate to the aspects, such as whether it is an essential food item or other fast moving consumer goods (FMCG) kind. After a detailed study of the product type and its nature, an effort should be made to estimate the cost of forecasting.

Sources of information need to be identified for data collection. Depending on the specific case, the data may be concerned with information such as:

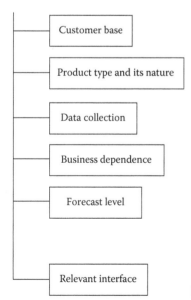

FIGURE 5.1   Influencing factors.

i. Geography
ii. Customer density
iii. Change in income level
iv. Liking/disliking, and any change pattern

For precise data collection and subsequent analysis, the effort level and the potential need for appropriate technology/expertise must be assessed. This aids in arriving at an appropriate/reasonable cost of forecasting.

In order to understand the business, its primary dependence should also be included where necessary. That is whether it is business to business, or business to consumers, among other aspects. Another factor is the forecast level. For example, it can be predicted at an aggregate level. Alternatively it may be concerned with the SKU level. In other words, a forecast might be for the product family or for the individual products. Such differentiation is used to ensure a suitable degree of efforts and eventually the forecasting expenditure.

For a perfect analysis, a relevant interface might become an area of concern. In order to solve a particular related operational issue, it is to be seen whether the interface is concerned with:

i. Supplier–Manufacturer
ii. Manufacturer–Warehouse
iii. Manufacturer–Dealer
iv. Warehouse–Distribution Centre
v. Wholesaler–Retailer
vi. Retailer–Customer

5.2   EVOLUTION STAGE

Most of the industrial and business organizations have a certain evolution stage in the present context. Such stages are represented by Figure 5.2.

In a rare case only, there might be absolutely no forecast relevance. However, in an initial stage, a firm may depend completely on the actual orders and a prediction of demand might not be felt necessary. But in most of the situations, one or few products or services may gain considerable attention or inclination of customers after crossing the mentioned initial stage. Thus, a forecast need may be felt by the entrepreneur or the concerned manager in the next stage of evolution.

After identification of the forecast need, a firm predicts the demand while implementing resources such as:

i. Information source for data collection
ii. Development of a forecasting system/process
iii. Relevant professional/practitioner

After an implementation of demand forecast in the decision making, the proposed plan is made operational for the purpose of buying and making, among other functions. However, the planned activity/performance needs to be reviewed if the

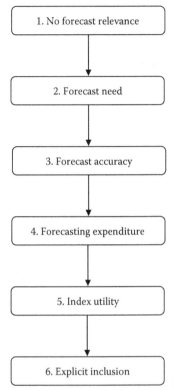

1. No forecast relevance

2. Forecast need

3. Forecast accuracy

4. Forecasting expenditure

5. Index utility

6. Explicit inclusion

FIGURE 5.2 Stages of evolution in the organization.

actual operation/performance varies. Thus, the forecast accuracy becomes the central attention in the next stage. Every effort is made to improve the forecast accuracy so that the mismatch between the planned and actual performance can be prevented to a reasonable extent.

Most of the progressive organizations are confined to this stage of forecast accuracy analysis and efforts to improve the methods. However, the present book also makes an attempt to link the forecasting expenditure with the potential benefit achieved by way of an improvement in the stated accuracy. In this fourth stage of evolution, any incremental enhancement in the relevant expenditure can be associated with the mentioned improvement in the present context of prediction.

After an emphasis on the forecasting expenditure, a suitable index can be utilized by the business firms in the next stage. Such an index reveals whether the proposed expenditure is justified in an overall context of the functions, such as:

i. Buying
ii. Making
iii. Priority planning

Finally, in the last stage, i.e., the sixth stage, a firm can consider the explicit inclusion of the forecasting expenditure in their regular planning process, such as the cyclic process among others. That is, it may appear along with other commonly known cost components in order to determine the total related cost. Subsequently, it is used in the formulation, optimization, and analysis in addition to other relevant practical applications. The important results are also summarized in the buying scenario now.

An optimal ordering quantity in each cycle is expressed as:

$$Q = \sqrt{\frac{2D(C + M)}{I}}$$

A minimized total cost is expressed as:

$$E = \sqrt{2DI(C + M)}$$

The making function is significant for relevant organizations. The important results are also summarized in the making scenario now.

An optimal production batch quantity in each cycle is expressed as:

$$Q = \sqrt{\frac{2D(C + M)}{I(1 - D/P)}}$$

A minimized total cost is expressed as:

$$E = \sqrt{2DI(C + M)(1 - D/P)}$$

In the context of priority planning or stock out inclusion, the expressions are developed for buying as well as making applications, such as:

a. Buying:

An optimal ordering quantity or lot size has been obtained as follows:

$$Q = \sqrt{\frac{2D(C + M)(K + I)}{IK}}$$

The optimal shortage or stock out quantities are obtained as:

$$J = \sqrt{\frac{2D(C + M)I}{K(K + I)}}$$

The total cost has been formulated as follows:

$$E = \sqrt{\frac{2D(C+M)IK}{(K+I)}}$$

(b) Making:

An optimal production batch quantity or lot size has been obtained as follows:

$$Q = \sqrt{\frac{2D(C+M)(I+K)}{IK(1-D/P)}}$$

The optimal shortage or stock out quantities are obtained as:

$$J = \sqrt{\frac{2D(C+M)I(1-D/P)}{K(K+I)}}$$

The total cost has been formulated as follows:

$$E = \sqrt{\frac{2D(C+M)IK(1-D/P)}{(K+I)}}$$

However, in the case of negligible forecast relevance or when the forecasting expenditure is ignored, the output parameters may also be modified and used accordingly. In certain cases, more than one option is available for the management to deal with a change. And it is possible to respond with a combination of parameters. For example, with setup cost reduction, the following options are available in order to have a similar production batch size:

i. Demand increase
ii. Holding cost reduction

Example 5.1:

In the making function, without a stock out scenario, an available set of information is as follows:
Setup cost, C = ₹45
Annual production rate, P = 960 units
Annual demand, D = 600 units
Annual inventory carrying cost per unit, I = ₹40

Production batch size may be expressed optimally as:

$$Q^* = \sqrt{\frac{2DC}{(1-D/P)I}}$$

$$= 60 \text{ units}$$

Now, if the setup cost is reduced by 20%, then:
$C_1 = ₹36$
The revised batch size is:
$Q = 53.67$
In order to have a similar batch size:

$$60 = \sqrt{\frac{2 \times D_1 \times 36}{(1 - D_1/960) \times 40}}$$

or

$D_1 = 648.65$ units

That is an 8.1% potential increase in demand.

However, if it is not possible to increase the demand by more than 5%, say, then the potential demand parameter now is 630 units. In order to achieve a similar batch size:

$$60 = \sqrt{\frac{2 \times 630 \times 36}{(1 - 630/960) I_1}}$$

or

$I_1 = 36.65$

That is an 8.36% reduction in holding cost.

Therefore, the objective can be achieved with a response concerning a combination of parameters if it is feasible.

For a general approach, let:
M = % reduction in setup cost
N = % increase in demand
R = % reduction in holding cost
Now:

$$C_1 = C\left(1 - \frac{M}{100}\right)$$

$$D_1 = D\left(1 + \frac{N}{100}\right)$$

$$I_1 = I\left(1 - \frac{R}{100}\right)$$

For similar batch size:

$$\sqrt{\frac{2DC}{(1 - D/P)I}} = \sqrt{\frac{2D_1 C_1}{(1 - D_1/P) I_1}}$$

or

$$DC(1 - D_1/P)I_1 = D_1C_1(1 - D/P)I$$

or

$$I_1 = \frac{D_1C_1(1 - D/P)I}{DC(1 - D_1/P)}$$

or

$$1 - \frac{R}{100} = \frac{D_1C_1(1 - D/P)}{DC(1 - D_1/P)}$$

or

$$\frac{R}{100} = \frac{DC(1 - D_1/P) - D_1C_1(1 - D/P)}{DC(1 - D_1/P)}$$

or

$$\frac{R}{100} = \frac{(1 - D_1/P) - (1 + N/100)(1 - M/100)(1 - D/P)}{(1 - D_1/P)}$$

or

$$\frac{R}{100} = \frac{1 - (1 + N/100)(D/P) - [(1 + N/100) - (D/P)(1 + N/100)](1 - M/100)}{1 - (1 + N/100)(D/P)}$$

or

$$\frac{R}{100} = \frac{\begin{array}{l} 1 - (1 + N/100)(D/P) - [(1 + N/100) - (D/P)(1 + N/100) \\ - (M/100)(1 + N/100) + (M/100)(D/P)(1 + N/100)] \end{array}}{1 - (1 + N/100)(D/P)}$$

or

$$\frac{R}{100} = \frac{\begin{array}{l} 1 - (1 + N/100)(D/P) - (1 + N/100) + (D/P)(1 + N/100) \\ + (M/100)(1 + N/100) - (M/100)(D/P)(1 + N/100) \end{array}}{1 - (1 + N/100)(D/P)}$$

or

$$\frac{R}{100} = \frac{1 - (1 + N/100) + (M/100)(1 + N/100) - (M/100)(D/P)(1 + N/100)}{1 - (1 + N/100)(D/P)}$$

or

$$\frac{R}{100} = \frac{(M/100)(1 + N/100) - (N/100) - (M/100)(D/P)(1 + N/100)}{1 - (1 + N/100)(D/P)}$$

or

$$R = \frac{M(1 + N/100) - N - (MD/P)(1 + N/100)}{1 - (1 + N/100)(D/P)}$$

Now:

Consider the present example. For M = 20:

$$R = \frac{7.5 - 0.925N}{0.375 - 0.00625N}$$

The values of R (for M = 20) are shown in Table 5.1, corresponding to the different values of M. Several combinations related to N and R can be generated. A suitable combination might be implemented, depending on its applicability concerning ease in altering the relevant parameters for specific value of M.

In addition to restoring the production batch size, there might be a need to restore the total related cost also.

TABLE 5.1

Values of R (Concerning I) Corresponding to N (Concerning D) for M = 20

S.No.	N	$R = \frac{7.5 - 0.925N}{0.375 - 0.00625N}$
1	1	17.83
2	2	15.59
3	3	13.26
4	4	10.86
5	5	8.36
6	6	5.78
7	7	3.09

In the context of restoring the batch size with explicit inclusion of forecasting expenditure, a suitable measure or combination can be adopted if it is feasible. Table 5.2 represents a practical guide for further analysis and application.

In the context of restoring the total related cost with explicit inclusion of forecasting expenditure, a suitable measure or combination can be adopted if it is feasible. Table 5.3 represents a practical guide for further analysis and application.

Example 5.2: In the making function with a stock out scenario, an available set of information is as follows:

Setup cost, C = ₹45
Annual production rate, P = 960 units

TABLE 5.2
A Practical Guide to Restore the Batch Size

Change Initiated by the Factor	Remedial Measure for Furthermore Analysis
A raise in demand	Setup cost decrease/Forecasting expenditure decrease/ Production rate increase
Demand reduction	Carrying cost decrease/Production rate reduction
Setup cost reduction/Forecasting expenditure decrease	A raise in demand/Carrying cost decrease/Production rate reduction
Setup cost increase	Forecasting expenditure decrease/Production rate increase
A raise in production rate	A raise in demand/Carrying cost decrease
Production rate reduction	Forecasting expenditure decrease/Setup cost decrease
Carrying cost increase	A raise in demand/Production rate decrease
Carrying cost reduction	Forecasting expenditure decrease/Setup cost decrease/ Production rate increase
Forecasting expenditure increase	Setup cost decrease/Production rate increase

TABLE 5.3
A Practical Guide to Restore the Total Related Cost

Change Initiated by the Factor	Remedial Measure for Furthermore Analysis
A raise in forecasting expenditure	Setup cost decrease/Carrying cost decrease/Production rate reduction
A raise in setup cost	Forecasting expenditure decrease/Carrying cost decrease/Production rate reduction
Carrying cost increase	Forecasting expenditure decrease/Setup cost decrease/Production rate reduction
Production rate increase	Forecasting expenditure/Setup/ Carrying cost decrease

Annual demand, D = 600 units
Annual inventory carrying cost per unit, I = ₹40
Annual shortage cost per unit, K = ₹100

Production batch size may be expressed optimally as:

$$Q^* = \sqrt{\frac{2DC\,(I + K)}{IK\,(1 - D/P)}}$$

$$= 71 \text{ units}$$

With setup cost reduction, the batch size also reduces. The objective of a similar production batch size can be achieved with a response concerning a combination of parameters if it is feasible.

For a general approach, let:

M = % reduction in setup cost
N = % increase in demand
R = % reduction in holding cost
Now:

$$C_1 = C\left(1 - \frac{M}{100}\right)$$

$$D_1 = D\left(1 + \frac{N}{100}\right)$$

$$I_1 = I\left(1 - \frac{R}{100}\right)$$

For similar batch size:

$$\sqrt{\frac{2DC\,(K + I)}{KI\,(1 - D/P)}} = \sqrt{\frac{2D_1 C_1\,(K + I_1)}{KI_1\,(1 - D_1/P)}}$$

or

$$\frac{DC\,(K + I)}{I\,(1 - D/P)} = \frac{D_1 C_1\,(K + I_1)}{I_1\,(1 - D_1/P)}$$

or

$$I_1\,(1 - D_1/P)DC\,(K + I) = I\,(1 - D/P)D_1 C_1\,(K + I_1)$$

or

$$I_1(1 - D_1/P)DC(K + I) = KI(1 - D/P)D_1C_1 + I_1I(1 - D/P)D_1C_1$$

or

$$I_1[(1 - D_1/P)DC(K + I) - I(1 - D/P)D_1C_1] = KI(1 - D/P)D_1C_1$$

or

$$I_1 = \frac{KI(1 - D/P)D_1C_1}{(1 - D_1/P)DC(K + I) - I(1 - D/P)D_1C_1}$$

or

$$1 - \frac{R}{100} = \frac{K(1 - D/P)D_1C_1}{(1 - D_1/P)DC(K + I) - I(1 - D/P)D_1C_1}$$

or

$$\frac{R}{100} = \frac{(1 - D_1/P)DC(K + I) - I(1 - D/P)D_1C_1 - K(1 - D/P)D_1C_1}{(1 - D_1/P)DC(K + I) - I(1 - D/P)D_1C_1}$$

or

$$\frac{R}{100} = \frac{[1 - (D/P)(1 + N/100)]DC(K + I) - (K + I)(1 - D/P)DC(1 + N/100)(1 - M/100)}{[1 - (D/P)(1 + N/100)]DC(K + I) - I(1 - D/P)DC(1 + N/100)(1 - M/100)}$$

or

$$\frac{R}{100} = \frac{[1 - (D/P)(1 + N/100)](K + I) - (K + I)(1 - D/P)(1 + N/100)(1 - M/100)}{[1 - (D/P)(1 + N/100)](K + I) - I(1 - D/P)(1 + N/100)(1 - M/100)}$$

or

$$\frac{R}{100} = \frac{\begin{aligned}(K + I)[1 - (D/P)(1 + N/100) \\ - \{(1 + N/100) - (D/P)(1 + N/100)\}(1 - M/100)]\end{aligned}}{\begin{aligned}K[1 - (D/P)(1 + N/100)] + I[1 - (D/P)(1 + N/100) \\ - \{(1 + N/100) - (D/P)(1 + N/100)\}(1 - M/100)]\end{aligned}}$$

or

$$\frac{R}{100} = \frac{\begin{aligned}(K + I)[1 - (D/P)(1 + N/100) - \{(1 + N/100) - (M/100)(1 + N/100) \\ + (M/100)(D/P)(1 + N/100)\}]\end{aligned}}{\begin{aligned}K[1 - (D/P)(1 + N/100)] + I[1 - (D/P)(1 + N/100) \\ - \{(1 + N/100) - (D/P)(1 + N/100) - (M/100)(1 + N/100) \\ + (M/100)(D/P)(1 + N/100)\}]\end{aligned}}$$

or

$$\frac{R}{100} = \frac{(K + I)[(M/100)(1 + N/100) - (N/100) - (M/100)(D/P)(1 + N/100)]}{\begin{aligned}K[1 - (D/P)(1 + N/100)] + I[(M/100)(1 + N/100) - (N/100) \\ - (M/100)(D/P)(1 + N/100)]\end{aligned}}$$

or

$$R = \frac{(K + I)[M(1 + N/100) - N - (MD/P)(1 + N/100)]}{K[1 - (D/P)(1 + N/100)] + (I/100)[M(1 + N/100) - N - (MD/P)(1 + N/100)]}$$

or

$$R = \frac{(K + I)[M(1 + N/100)(1 - D/P) - N]}{K[1 - (D/P)(1 + N/100)] + (I/100)[M(1 + N/100)(1 - D/P) - N]}$$

Consider the present example. For M = 20:

$$R = \frac{1050 - 129.5N}{40.5 - 0.995N}$$

The values of R (for M = 20) are shown in Table 5.4, corresponding to the different values of M. Several combinations for N and R can be generated. A suitable combination might be implemented, depending on its applicability concerning ease in altering the relevant parameters for specific value of M.

The analysis can also be extended in multiple ways, depending on the evolution stage, i.e., with the explicit inclusion of forecasting efforts and an associated expenditure.

5.3 BUSINESS USE

Logistics play an important role in the local or global business. Distribution centers may be situated in a country or globally. Depending on the location and allied

TABLE 5.4

Values of R (I) Corresponding to N (D) with Stock Out for M = 20

S.No.	N	$R = \frac{1050 - 129.5N}{40.5 - 0.995N}$
1	1	23.30
2	2	20.54
3	3	17.63
4	4	14.57
5	5	11.33
6	6	7.91
7	7	4.28

resources, the forecasting expenditure varies and, thus, its estimation should incorporate relevant efforts. Relevant investments and efforts would lead to a more precise assessment of a corresponding expenditure on the forecast. This also helps in subsequent analysis to be more accurate and implementable in real-life situations. Such an implementation is expected to result in a relatively beneficial scenario either within a particular country or in the context of global spread of the manufacturing, distribution, and allied commercial activities.

Among other aspects, business use can also be differentiated on the basis of interfaces. Such an interface might be relevant between a seller firm and a buyer firm, as shown in Figure 5.3.

Forecasting efforts on the part of a particular buyer firm can affect the seller firm in one way or the other in an overall business scenario where such a scenario may relate to a channel case or an independent firm. Depending on the specific case, a particular buyer firm might also be asked to enhance the corresponding forecast expenditure. However, such a decision is dependent on the proposed threshold value analysis in the present book, i.e., when it makes business sense to increase the expenditure for forecasting efforts.

Similarly, another interface can concern a producing firm and warehouse, as represented by Figure 5.4.

FIGURE 5.3 Interface between buyer and seller firm.

FIGURE 5.4 Interface between producer and warehouse.

FIGURE 5.5   Interface between warehouse and distribution center.

Also, it can relate to a central warehouse and distribution center, as shown in Figure 5.5.

In general, such an interface is between an upstream and downstream firm. The specific case may be considered pertaining to a single/multiple firm(s). The application also corresponds to the situations when it is an independent firm or in a channel relationship. The business use depends on the key functions of whether it is buying or making because the relevant threshold value takes into consideration the parameters accordingly. With the help of computational/analytical efforts, it would be easier to know whether a particular firm should increase the forecasting expenditure. Such an increase is expected to reap an overall business/commercial benefit in multiple ways for a wide variety of organizations.

Appendix

In the present context, cycle time refers to the duration in which a certain number of items or quantity is produced and consumed completely. After this, a similar cycle repeats itself. Inventory planning is also emphasized, considering the importance of such a cycle time.

A.1 IMPORTANCE OF CYCLE TIME

As shown in Figure A.1, the stock buildup rate is (P – D), where P is the production rate and D is the demand rate. Stock increases during the production time. At the end of the production time, stock starts decreasing. As soon as the stock level becomes zero, the next similar production cycle begins.

Cycle time (T) is required to be reduced with reference to a certain optimization when there is:

   i. Reduced facility setup cost
   ii. Increased production rate
   iii. Increased inventory holding cost
   iv. Reduced demand

Furthermore, when shelf life is an issue and the product should be consumed timely, the production cycle time might be reduced. These factors are also shown in Figure A.2.

Cycle time (C.T.) may be increased (Figure A.3) when there is:

   i. Increased facility setup cost
   ii. Reduced production rate
   iii. Reduced inventory holding cost
   iv. Increased demand

Variation in the cycle time (T), i.e., a reduction or increase, needs to be studied. For this purpose, the problem should be formulated in terms of T.

Now, one cycle of production is shown in Figure A.4.

For example, if annual demand (D) is 12,000 products and the cycle time is 0.2 year, then the production quantity per setup is:

$12{,}000 \times 0.2 = 2{,}400$

or

$1{,}200 \times \frac{1}{5} = 2{,}400$, because there are five production cycles in one year.

In order to generalize, production batch size = DT.

Production time (P.T.) = $\frac{DT}{P}$

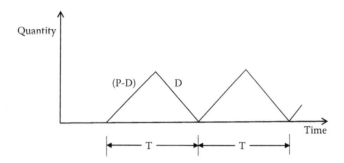

FIGURE A.1   Cycle time.

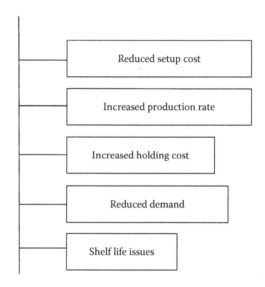

FIGURE A.2   Factors for a reduced cycle time.

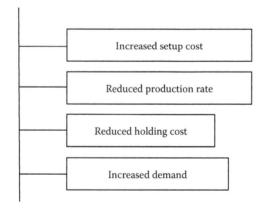

FIGURE A.3   Factors for an increased cycle time.

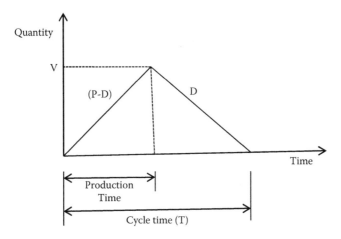

FIGURE A.4   A cycle of production.

P.T. can also be expressed as:

$$\frac{V}{(P-D)}$$

where V = maximum stock during the cycle.
    Now:

$$\frac{V}{(P-D)} = \frac{DT}{P}$$

or

$$V = \frac{DT(P-D)}{P}$$

or

$$V = DT(1-D/P)$$

Because average stock is $\left(\frac{V}{2}\right)$, the annual inventory holding cost is:

$$\text{AIC} = \frac{V}{2.}\cdot I$$

or

$$\text{AIC} = \frac{DTI(1-D/P)}{2} \qquad\qquad\qquad (\text{A.1})$$

If cycle time (T) is 0.2 year, then the number of facility setups in one year is:

$$\frac{1}{0.2} = 5$$

In order to generalize, the number of setups in one year is:

$$\frac{1}{T}$$

and the annual production setup cost is as follows:

$$APC = \frac{1}{T}.\, C = \frac{C}{T} \tag{A.2}$$

While adding Eqs. (A.1) and (A.2), the total annual cost is:

$$E = AIC + APC$$

or

$$E = \frac{DTI\,(1 - D/P)}{2} + \frac{C}{T} \tag{A.3}$$

In order to obtain the optimal value of T, differentiate w.r.t. T and equate to zero:

$$\frac{DI\,(1 - D/P)}{2} - \frac{C}{T^2} = 0$$

or

$$\frac{DI\,(1 - D/P)}{2} = \frac{C}{T^2}$$

or

$$T^2 = \frac{2C}{DI\,(1 - D/P)}$$

or

$$T* = \sqrt{\frac{2C}{DI\,(1 - D/P)}} \tag{A.4}$$

For the optimal value of E, substitute the value of DI(1 − D/P) from Eq. (A.4) into Eq. (A.3).

$$E^* = \frac{2C}{T^2} \cdot \frac{T}{2} + \frac{C}{T}$$

or

$$E^* = \frac{C}{T} + \frac{C}{T}$$

or

$$E^* = \frac{2C}{T} \tag{A.5}$$

Example A.1: Consider the following parameters:

Annual demand, D = 600
Annual inventory holding cost per unit, I = ₹35
Annual production rate, P = 960
Setup cost, C = ₹50

From (A.4), cycle time is:

$$T = \sqrt{\frac{2 \times 50}{600 \times 35(1 - 600/960)}}$$

or
T = 0.1127 year ≈ 0.113 year
From (A.5), the total cost is:

$$E = \frac{2 \times 50}{0.1127}$$

or
E = ₹887.41

A.2   VARIATION IN CYCLE TIME

Variation in cycle time can be on the lower or higher side.

A.2.1 LOWER CYCLE TIME

Optimal cycle time decreases because of:

    i. Reduced facility setup cost
   ii. Increased production rate
  iii. Increased holding cost
  iv. Decreased demand

Now, let:
   M = % variation in parameter under consideration
   T_1 = Reduced cycle time

   i. Reduced setup cost:

Since the parameter under consideration is setup cost, M refers to the % reduction in setup cost.
   Now:

$$C_1 = \left(1 - \frac{M}{100}\right)C$$

Decrease in the cycle time $= T - T_1$

$$= \sqrt{\frac{2C}{DI(1 - D/P)}} - \sqrt{\frac{2C_1}{DI(1 - D/P)}}$$

$$= \sqrt{\frac{2C}{DI(1 - D/P)}} \left[1 - \sqrt{\frac{C_1}{C}}\right]$$

$$= \sqrt{\frac{2C}{DI(1 - D/P)}} \left[1 - \sqrt{\left(1 - \frac{M}{100}\right)}\right]$$

And:

$$\% \text{ decrease in C.T.} = 1 - \sqrt{\left(1 - \frac{M}{100}\right)}$$

In order to illustrate, consider the parameters of Example A.1 as follows:

D	C	I	P	T	E
600	50	35	960	0.113	887.41

After implementation of the reduced setup cost, the approximate effects are given as follows:

% Decrease in C	5%	10%	15%	20%	25%
C	47.50	45.00	42.50	40.00	37.50
T	0.110	0.107	0.104	0.101	0.098
% Decrease in T	2.53%	5.13%	7.80%	10.56%	13.40%
E	864.94	841.87	818.15	793.73	768.52
% Decrease in E	2.53%	5.13%	7.80%	10.56%	13.40%

Percent reduction in the cycle time is lower than that in the facility setup cost.

ii. Increased production rate:

$$P_1 = \left(1 + \frac{M}{100}\right)P$$

Decrease in cycle time $= T - T_1$

$$= \sqrt{\frac{2C}{DI(1 - D/P)}} - \sqrt{\frac{2C}{DI(1 - D/P_1)}}$$

$$= \sqrt{\frac{2C}{DI(1 - D/P)}}\left[1 - \sqrt{\frac{(1 - D/P)}{(1 - D/P_1)}}\right]$$

$$= \sqrt{\frac{2C}{DI(1 - D/P)}}\left[1 - \sqrt{\frac{(1 - D/P)}{1 - \{D/P(1 + M/100)\}}}\right]$$

and:

$$\% \text{ decrease in C.T.} = 1 - \sqrt{\frac{(1 - D/P)}{1 - \{D/P(1 + M/100)\}}}$$

In order to illustrate, consider the reference parameters as follows:

D	C	I	P	T	E
600	50	35	960	0.113	887.41

After implementation of the production rate, the effects in terms of approximate values are given as follows:

% Increase in P	5%	10%	15%	20%	25%	30%
P	1008	1056	1104	1152	1200	1248

(Continued)

T	0.108	0.105	0.102	0.100	0.098	0.096
% Decrease in T	3.75%	6.81%	9.37%	11.53%	13.40%	15.02%
E	921.95	952.27	979.13	1003.12	1024.70	1044.21
% Increase in E	3.89%	7.31%	10.34%	13.04%	15.47%	17.67%

Percent decrease in the cycle time is lower in comparison with the % increase in the production rate.

iii. Increased holding cost:

$$I_1 = \left(1 + \frac{M}{100}\right)I$$

Decrease in cycle time $= T - T_1$

$$= \sqrt{\frac{2C}{DI(1 - D/P)}} - \sqrt{\frac{2C}{DI_1(1 - D/P)}}$$

$$= \sqrt{\frac{2C}{DI(1 - D/P)}}\left[1 - \sqrt{\frac{I}{I_1}}\right]$$

$$= \sqrt{\frac{2C}{DI(1 - D/P)}}\left[1 - \sqrt{\frac{1}{(1 + M/100)}}\right]$$

and:

$$\% \text{ decrease in C.T.} = 1 - \sqrt{\frac{1}{(1 + M/100)}}$$

In order to illustrate, consider the reference parameters as follows:

D	C	I	P	T	E
600	50	35	960	0.113	887.41

After implementation of the increased holding cost, the effects in terms of approximate values are provided below:

% Increase in I	5%	10%	15%	20%	25%
I	36.75	38.50	40.25	42.00	43.75
T	0.110	0.107	0.105	0.103	0.101
% Increase in T	2.41%	4.65%	6.75%	8.71%	10.56%
E	909.33	930.73	951.64	972.11	992.16
% Increase in E	2.47%	4.88%	7.24%	9.54%	11.80%

iv. Decreased demand:

$$D_1 = \left(1 - \frac{M}{100}\right)D$$

Decrease in cycle time $= T - T_1$

$$= \sqrt{\frac{2C}{DI(1-D/P)}} - \sqrt{\frac{2C}{D_1 I(1-D_1/P)}}$$

$$= \sqrt{\frac{2C}{DI(1-D/P)}}\left[1 - \sqrt{\frac{D(1-D/P)}{D_1(1-D_1/P)}}\right]$$

$$= \sqrt{\frac{2C}{DI(1-D/P)}}\left[1 - \sqrt{\frac{(1-D/P)}{(1-M/100)\{1-D(1-M/100)/P\}}}\right]$$

and:

$$\% \text{ decrease in C.T.} = 1 - \sqrt{\frac{(1-D/P)}{(1-M/100)\{1-D(1-M/100)/P\}}}$$

The reference parameters are as follows:

D	C	I	P	T	E
600	50	35	960	0.113	887.41

When demand is decreased in this example, the effects in terms of approximate values are given below:

% Decrease in D	5%	10%	15%	20%
D	570.00	540.00	510.00	480.00
T	0.111	0.110	0.109	0.109
% Decrease in T	1.43%	2.41%	2.99%	3.18%
E	900.26	909.33	914.72	916.52
% Increase in E	1.45%	2.47%	3.08%	3.28%

Table A.1 summarizes the generalized results.

TABLE A.1
Results for Reduced Cycle Time

Variation in Input Parameter	Decrease in Optimal C.T.

Reduced facility setup cost

$\text{Decrease in the cycle time} = \sqrt{\dfrac{2C}{DI(1-D/P)}}\left[1 - \sqrt{\left(1 - \dfrac{M}{100}\right)}\right]$ % decrease in

$\text{C.T.} = 1 - \sqrt{\left(1 - \dfrac{M}{100}\right)}$

Increased production rate

$\text{Decrease in the cycle time} = \sqrt{\dfrac{2C}{DI(1-D/P)}}\left[1 - \sqrt{\dfrac{(1-D/P)}{1-\{D/P(1+M/100)\}}}\right]$ %

$\text{decrease in C.T.} = 1 - \sqrt{\dfrac{(1-D/P)}{1-\{D/P(1+M/100)\}}}$

Increased holding cost

$\text{Decrease in the cycle time} = \sqrt{\dfrac{2C}{DI(1-D/P)}}\left[1 - \sqrt{\dfrac{1}{(1+M/100)}}\right]$ % decrease in

$\text{C.T.} = 1 - \sqrt{\dfrac{1}{(1+M/100)}}$

Decreased demand

$\text{Decrease in the cycle time} =$

$\sqrt{\dfrac{2C}{DI(1-D/P)}}\left[1 - \sqrt{\dfrac{(1-D/P)}{(1-M/100)\{1-D(1-M/100)/P\}}}\right]$ % decrease in C.T. $=$

$1 - \sqrt{\dfrac{(1-D/P)}{(1-M/100)\{1-D(1-M/100)/P\}}}$

A.2.2 HIGHER CYCLE TIME

Optimal cycle time increases because of:

  i. Increased facility setup cost
 ii. Decreased production rate
iii. Decreased holding cost
 iv. Increased demand

Now:
   M = % variation in parameters under consideration
   T_1 = Increased cycle time

   i. Increased setup cost:

$$C_1 = \left(1 + \frac{M}{100}\right)C$$

Increase in the cycle time $= T_1 - T$

$$= \sqrt{\frac{2C_1}{DI(1-D/P)}} - \sqrt{\frac{2C}{DI(1-D/P)}}$$

$$= \sqrt{\frac{2C}{DI(1-D/P)}} \left[\sqrt{\frac{C_1}{C}} - 1 \right]$$

$$= \sqrt{\frac{2C}{DI(1-D/P)}} \left[\sqrt{\left(1 + \frac{M}{100}\right)} - 1 \right]$$

and:

$$\% \text{ increase in C.T.} = \sqrt{\left(1 + \frac{M}{100}\right)} - 1$$

In order to illustrate, consider the reference parameters as follows:

D	C	I	P	T	E
600	50	35	960	0.113	887.41

After implementation of the increased setup cost, the approximate effects are given as follows:

% Increase in C	5%	10%	15%	20%	25%
C	52.50	55.00	57.50	60.00	62.50
T	0.115	0.118	0.121	0.123	0.126
% Increase in T	2.47%	4.88%	7.24%	9.54%	11.80%
E	909.33	930.73	951.64	972.11	992.16
% Increase in E	2.47%	4.88%	7.24%	9.54%	11.80%

ii. Decreased production rate:

$$P_1 = \left(1 - \frac{M}{100}\right)P$$

Increase in the cycle time $= T_1 - T$

$$= \sqrt{\frac{2C}{DI(1-D/P_1)}} - \sqrt{\frac{2C}{DI(1-D/P)}}$$

$$= \sqrt{\frac{2C}{DI(1-D/P)}} \left[\sqrt{\frac{(1-D/P)}{(1-D/P_1)}} - 1 \right]$$

$$= \sqrt{\frac{2C}{DI(1-D/P)}} \left[\sqrt{\frac{(1-D/P)}{1-\{D/P(1-M/100)\}}} - 1 \right]$$

and:

$$\% \text{ increase in C.T.} = \sqrt{\frac{(1 - D/P)}{1 - \{D/P(1 - M/100)\}}} - 1$$

The reference parameters are as follows:

D	C	I	P	T	E
600	50	35	960	0.113	887.41

The approximate effects are given below:

% Decrease in P	5%	10%	15%	20%	25%	30%
P	912	864	816	768	720	672
T	0.118	0.125	0.134	0.148	0.169	0.211
% Increase in T	4.70%	10.78%	19.02%	30.93%	50.00%	87.08%
E	847.60	801.04	745.58	677.77	591.61	474.34
% Decrease in E	4.49%	9.73%	15.98%	23.62%	33.33%	46.55%

iii. Decreased holding cost:

$$I_1 = \left(1 - \frac{M}{100}\right)I$$

Increase in the cycle time $= T_1 - T$

$$= \sqrt{\frac{2C}{DI_1(1 - D/P)}} - \sqrt{\frac{2C}{DI(1 - D/P)}}$$

$$= \sqrt{\frac{2C}{DI(1 - D/P)}} \left[\sqrt{\frac{I}{I_1}} - 1\right]$$

$$= \sqrt{\frac{2C}{DI(1 - D/P)}} \left[\sqrt{\frac{1}{(1 - M/100)}} - 1\right]$$

and:

$$\% \text{ increase in C.T.} = \sqrt{\frac{1}{(1 - M/100)}} - 1$$

In order to illustrate, consider the reference parameters as before:

D	C	I	P	T	E
600	50	35	960	0.113	887.41

   After implementation of the decreased holding cost, the effects in terms of approximate values are provided below:

% Decrease in I	5%	10%	15%	20%	25%
I	33.25	31.50	29.75	28.00	26.25
T	0.116	0.119	0.122	0.126	0.130
% Increase in T	2.60%	5.41%	8.47%	11.80%	15.47%
E	864.94	841.87	818.15	793.73	768.52
% Decrease in E	2.53%	5.13%	7.80%	10.56%	13.40%

   iv. Increased demand:

$$D_1 = \left(1 + \frac{M}{100}\right)D$$

Increase in the cycle time $= T_1 - T$

$$= \sqrt{\frac{2C}{D_1 I (1 - D_1/P)}} - \sqrt{\frac{2C}{DI(1 - D/P)}}$$

$$= \sqrt{\frac{2C}{DI(1 - D/P)}} \left[\sqrt{\frac{D(1 - D/P)}{D_1(1 - D_1/P)}} - 1\right]$$

$$= \sqrt{\frac{2C}{DI(1 - D/P)}} \left[\sqrt{\frac{(1 - D/P)}{(1 + M/100)\{1 - D(1 + M/100)/P\}}} - 1\right]$$

and:

$$\% \text{ increase in C.T.} = \sqrt{\frac{(1 - D/P)}{(1 + M/100)\{1 - D(1 + M/100)/P\}}} - 1$$

The reference parameters are as follows:

D	C	I	P	T	E
600	50	35	960	0.113	887.41

When demand is increased in this example, the effects in terms of approximate values are given below:

% Increase in D	5%	10%	15%	20%	25%
D	630.00	660.00	690.00	720.00	750.00
T	0.115	0.118	0.121	0126	0.132
% Increase in T	1.93%	4.45%	7.68%	11.80%	17.11%
E	870.61	849.63	824.15	793.73	757.77
% Increase in E	1.89%	4.26%	7.13%	10.56%	14.61%

Table A.2 summarizes the generalized results.

A.3 BACKORDERING SITUATION

A cycle of production with shortages that are completely backordered is shown in Figure A.5.

Maximum shortage quantity = J

TABLE A.2

Results for Increased Cycle Time

Variation in Input Parameters	Increase in Optimal C.T.
Increased facility setup cost	Increase in the cycle time $= \sqrt{\dfrac{2C}{DI(1-D/P)}}\left[\sqrt{\left(1+\dfrac{M}{100}\right)}-1\right]$ % increase in C.T. $= \sqrt{\left(1+\dfrac{M}{100}\right)}-1$
Decreased production rate	Increase in the cycle time $= \sqrt{\dfrac{2C}{DI(1-D/P)}}\left[\sqrt{\dfrac{(1-D/P)}{1-\{D/P(1-M/100)\}}}-1\right]$ % increase in C.T. $= \sqrt{\dfrac{(1-D/P)}{1-\{D/P(1-M/100)\}}}-1$
Decreased holding cost	Increase in the cycle time $= \sqrt{\dfrac{2C}{DI(1-D/P)}}\left[\sqrt{\dfrac{1}{(1-M/100)}}-1\right]$ % increase in C.T. $= \sqrt{\dfrac{1}{(1-M/100)}}-1$
Increased demand	Increase in the cycle time $=$ $\sqrt{\dfrac{2C}{DI(1-D/P)}}\left[\sqrt{\dfrac{(1-D/P)}{(1+M/100)\{1-D(1+M/100)/P\}}}-1\right]$ % increase in C.T. $= \sqrt{\dfrac{(1-D/P)}{(1+M/100)\{1-D(1+M/100)/P\}}}-1$

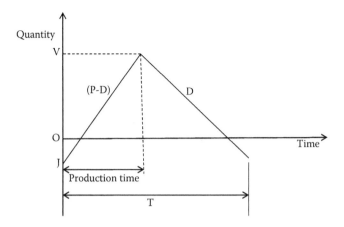

FIGURE A.5 A cycle of production with shortages.

$$\text{Period during which shortages occur} = \frac{J}{(P-D)} + \frac{J}{D} = \frac{J}{D(1-D/P)}$$

$$\text{Average shortage quantity} = \frac{J}{2}$$

Since there are $\frac{1}{T}$ number of cycles in a year, the annual shortage cost is:

$$\text{ASC} = \frac{J}{2} \cdot \frac{J}{D(1-D/P)} \cdot \frac{1}{T} \cdot K$$

where K = Annual shortage cost per unit.
   Or:

$$\text{ASC} = \frac{J^2 K}{2TD(1-D/P)} \qquad (A.6)$$

As the production batch size is (DT),

$$\text{Production time} = \frac{DT}{P}$$

However, the production time can also be expressed as:

$$\frac{(V+J)}{(P-D)}$$

Therefore:

$$\frac{(V + J)}{(P - D)} = \frac{DT}{P}$$

or

$$V = \frac{DT\,(P - D)}{P} - J$$

or

$$V = DT\,(1 - D/P) - J \qquad\qquad (A.7)$$

Positive stock exists during the period:

$$\frac{V}{(P - D)} + \frac{V}{D}$$

$$= \frac{V}{D\,(1 - D/P)}$$

As there are $\frac{1}{T}$ number of cycles in a year, the annual inventory holding cost is given as:

$$AIC = \frac{V}{2} \cdot \frac{V}{D\,(1 - D/P)} \cdot \frac{1}{T} \cdot I$$

or

$$AIC = \frac{V^2 I}{2DT\,(1 - D/P)}$$

Substituting the value of V from Eq. (A.7),

$$AIC = \frac{[DT\,(1 - D/P) - J]^2 I}{2DT\,(1 - D/P)}$$
$$= \frac{I\,[D^2 T^2\,(1 - D/P)^2 - 2DTJ\,(1 - D/P) + J^2]}{2DT\,(1 - D/P)}$$

or

$$AIC = \frac{IDT\,(1 - D/P)}{2} - IJ + \frac{IJ^2}{2DT\,(1 - D/P)} \qquad\qquad (A.8)$$

Annual production setup cost is:

$$\text{APC} = \frac{C}{T} \tag{A.9}$$

By adding Eqs. (A.6), (A.8), and (A.9), the total annual cost is:

$$E = \text{ASC} + \text{AIC} + \text{APC}$$

or

$$E = \frac{J^2 K}{2TD(1 - D/P)} + \frac{IDT(1 - D/P)}{2} - IJ + \frac{IJ^2}{2DT(1 - D/P)} + \frac{C}{T} \tag{A.10}$$

Differentiating partially w.r.t. J and equating to zero:

$$\frac{JK}{TD(1 - D/P)} - I + \frac{IJ}{DT(1 - D/P)} = 0$$

or

$$\frac{J\,(K + I)}{TD(1 - D/P)} = I$$

or

$$J = \frac{IDT(1 - D/P)}{(K + I)} \tag{A.11}$$

Substituting the value of J in Eq. (A.10):

$$E = \frac{IDT(1 - D/P)}{2} - \frac{I^2 DT(1 - D/P)}{2(K + I)} + \frac{C}{T} \tag{A.12}$$

In order to obtain the optimal value of T, differentiate with respect to T and equate to zero:

$$\frac{ID(1 - D/P)}{2} - \frac{I^2 D(1 - D/P)}{2(K + I)} - \frac{C}{T^2} = 0$$

or

$$\frac{C}{T^2} = \frac{ID(1 - D/P)[(K + I) - I]}{2(K + I)}$$

or

$$T^2 = \frac{2C(K+I)}{KID(1-D/P)}$$

or

$$T^* = \sqrt{\frac{2C(K+I)}{KID(1-D/P)}} \qquad (A.13)$$

From Eq. (A.13):

$$\frac{DI(1-D/P)}{2} = \frac{C(K+I)}{KT^2}$$

Putting this value in Eq. (A.12),

$$E^* = T\left[\frac{C(K+I)}{KT^2}\right] - \frac{TI}{(K+I)}\left[\frac{C(K+I)}{KT^2}\right] + \frac{C}{T}$$

or

$$E^* = \frac{C(K+I)}{KT} - \frac{IC}{KT} + \frac{C}{T}$$

or

$$E^* = \frac{C}{T} + \frac{C}{T}$$

or

$$E^* = \frac{2C}{T} \qquad (A.14)$$

Example A.2: Consider the following parameters:

Annual demand, D = 600
Annual inventory holding cost per unit, I = ₹35
Annual production rate, P = 960
Setup cost, C = ₹50
Annual shortage cost per unit = ₹100

From Eq. (A.13), cycle time is:

$$T = \sqrt{\frac{2 \times 50 \times (100 + 35)}{600 \times 35 \times 100 \times (1 - 600/960)}}$$

or
T = 0.1309 year ≈ 0.131 year
From Eq. (A.14), the total cost is:

$$E = \frac{2 \times 50}{0.1309}$$

or
E = ₹ 763.76

Compared to Example A.1, the optimal cycle time is higher in order to incorporate the shortages. However, the overall total cost is lower.

A.3.1   DOWNWARD VARIATION IN CYCLE TIME

Optimal cycle time decreases because of:

i. Reduced facility setup cost
ii. Increased production rate
iii. Increased holding cost
iv. Decreased demand
v. Increased shortage cost

Now:
M = % variation in parameter under consideration
T_1 = Reduced cycle time

i. Reduced setup cost:

Since the parameter under consideration is setup cost, M refers to the % reduction in setup cost.
Now:

$$C_1 = \left(1 - \frac{M}{100}\right)C$$

Decrease in the cycle time = $T - T_1$

$$= \sqrt{\frac{2C(K+I)}{KID(1-D/P)}} - \sqrt{\frac{2C_1(K+I)}{KID(1-D/P)}}$$

$$= \sqrt{\frac{2C(K+I)}{KID(1-D/P)}}\left[1 - \sqrt{\frac{C_1}{C}}\right]$$

$$= \sqrt{\frac{2C(K+I)}{KID(1-D/P)}}\left[1 - \sqrt{\left(1 - \frac{M}{100}\right)}\right]$$

and:

$$\% \text{ decrease in C.T.} = 1 - \sqrt{\left(1 - \frac{M}{100}\right)}$$

In order to illustrate, consider the parameters of Example A.2 as follows:

D	C	I	P	K	T	E
600	50	35	960	100	0.131	763.76

After implementation of the reduced setup cost, the approximate effects are given as follows:

% Decrease in C	5%	10%	15%	20%	25%	30%
C	47.50	45.00	42.50	40.00	37.50	35.00
T	0.128	0.124	0.121	0.117	0.113	0.110
% Decrease in T	2.53%	5.13%	7.80%	10.56%	13.40%	16.33%
E	744.42	724.57	704.15	683.13	661.44	639.01
% Decrease in E	2.53%	5.13%	7.80%	10.56%	13.40%	16.33%

ii. Increased production rate:

$$P_1 = \left(1 + \frac{M}{100}\right)P$$

Decrease in the cycle time $= T - T1$

$$= \sqrt{\frac{2C(K+I)}{KID(1-D/P)}} - \sqrt{\frac{2C(K+I)}{KID(1-D/P_1)}}$$

$$= \sqrt{\frac{2C(K+I)}{KID(1-D/P)}}\left[1 - \sqrt{\frac{(1-D/P)}{(1-D/P_1)}}\right]$$

$$= \sqrt{\frac{2C(K+I)}{KID(1-D/P)}}\left[1 - \sqrt{\frac{(1-D/P)}{1-\{D/P(1+M/100)\}}}\right]$$

and:

$$\% \text{ decrease in C.T.} = 1 - \sqrt{\frac{(1-D/P)}{1-\{D/P(1+M/100)\}}}$$

In order to illustrate, consider the reference parameters as follows:

D	C	I	P	K	T	E
600	50	35	960	100	0.131	763.76

After implementation of the increased production rate, the effects in terms of approximate values are given as follows:

% Increase in P	5%	10%	15%	20%	25%	30%
P	1008	1056	1104	1152	1200	1248
T	0.126	0.122	0.119	0.116	0.113	0.111
% Decrease in T	3.75%	6.81%	9.37%	11.53%	13.40%	15.02%
E	793.49	819.58	842.70	863.35	881.92	898.72
% Increase in E	3.89%	7.31%	10.34%	13.04%	15.47%	17.67%

Percent decrease in the cycle time is lower in comparison with the % increase in the production rate.

$$I_1 = \left(1 + \frac{M}{100}\right)I$$

Decrease in the cycle time $= T - T_1$

$$= \sqrt{\frac{2C(K+I)}{KID(1-D/P)}} - \sqrt{\frac{2C(K+I_1)}{KI_1D(1-D/P)}}$$

$$= \sqrt{\frac{2C(K+I)}{KID(1-D/P)}}\left[1 - \sqrt{\frac{I(K+I_1)}{I_1(K+I)}}\right]$$

$$= \sqrt{\frac{2C(K+I)}{KID(1-D/P)}}\left[1 - \sqrt{\frac{K+I(1+M/100)}{(1+M/100)(K+I)}}\right]$$

iii. Increased holding cost:

$$1 - \sqrt{\frac{K+I(1+M/100)}{(1+M/100)(K+I)}}$$

In order to illustrate, consider the reference parameters as before:

D	C	I	P	K	T	E
600	50	35	960	100	0.131	763.76

After implementation of the increased holding cost, the effects in terms of approximate values are provided as follows:

% Increase in I	5%	10%	15%	20%	25%	30%
I	36.75	38.50	40.25	42.00	43.75	45.50
T	0.128	0125	0.122	0.120	0.117	0.115
% Decrease in T	2.41%	4.65%	6.75%	8.71%	10.56%	12.29%
E	782.62	801.04	819.04	836.66	853.91	870.82
% Increase in E	2.47%	4.88%	7.24%	9.54%	11.80%	14.02%

iv. Decreased demand:

$$D_1 = \left(1 - \frac{M}{100}\right)D$$

Decrease in the cycle time $= T - T_1$

$$= \sqrt{\frac{2C(K+I)}{KID(1-D/P)}} - \sqrt{\frac{2C(K+I)}{KID_1(1-D_1/P)}}$$

$$= \sqrt{\frac{2C(K+I)}{KID(1-D/P)}}\left[1 - \sqrt{\frac{D(1-D/P)}{D_1(1-D_1/P)}}\right]$$

$$= \sqrt{\frac{2C(K+I)}{KID(1-D/P)}}\left[1 - \sqrt{\frac{(1-D/P)}{(1-M/100)\{1-D(1-M/100)/P\}}}\right]$$

and:

$$\% \text{ decrease in C.T.} = 1 - \sqrt{\frac{(1-D/P)}{(1-M/100)\{1-D(1-M/100)/P\}}}$$

The reference parameters are as follows:

D	C	I	P	K	T	E
600	50	35	960	100	0.131	763.76

When demand is decreased in this example, the effects in terms of approximate values are given below:

% Decrease in D	5%	10%	15%	20%
D	570.00	540.00	510.00	480.00
T	0.129	0.128	0.127	0.127

% Decrease in T	1.53	2.29	3.05	3.05
E	774.82	782.62	787.27	788.81
% Increase in E	1.45%	2.47%	3.08%	3.28%

v. Increased shortage cost:

$$K_1 = \left(1 + \frac{M}{100}\right)K$$

Decrease in the cycle time $= T - T_1$

$$= \sqrt{\frac{2C(K+I)}{KID(1-D/P)}} - \sqrt{\frac{2C(K_1+I)}{K_1ID(1-D/P)}}$$

$$= \sqrt{\frac{2C(K+I)}{KID(1-D/P)}} \left[1 - \sqrt{\frac{K(K_1+I)}{K_1(K+I)}}\right]$$

$$= \sqrt{\frac{2C(K+I)}{KID(1-D/P)}} \left[1 - \sqrt{\frac{I+K(1+M/100)}{(1+M/100)(K+I)}}\right]$$

and:

$$\% \text{ decrease in C.T.} = 1 - \sqrt{\frac{I+K(1+M/100)}{(1+M/100)(K+I)}}$$

The reference parameters are as follows:

D	C	I	P	K	T	E
600	50	35	960	100	0.131	763.76

The effects in terms of approximate values are given below:

% Increase in K	5%	10%	15%	20%	25%	30%
K	105	110	115	120	125	130
T	0.1301	0.1294	0.1287	0.1281	0.1275	0.1270
% Decrease in T	0.62%	1.19%	1.71%	2.18%	2.63%	3.04%
E	768.52	772.93	777.01	780.82	784.37	787.69
% Increase in E	0.62%	1.20%	1.73%	2.23%	2.70%	3.13%

Table A.3 summarizes the generalized results.

TABLE A.3

Results for the Reduced Cycle Time with Shortages

Variation in Input Parameter	Decrease in Optimal C.T.
Reduced facility setup cost	Decrease in the cycle time = $\sqrt{\dfrac{2C(K+I)}{KID(1-D/P)}}\left[1-\sqrt{\left(1-\dfrac{M}{100}\right)}\right]$ % decrease in C.T. $= 1 - \sqrt{\left(1 - \dfrac{M}{100}\right)}$
Increased production rate	Decrease in the cycle time = $\sqrt{\dfrac{2C(K+I)}{KID(1-D/P)}}\left[1-\sqrt{\dfrac{(1-D/P)}{1-\{D/P(1+M/100)\}}}\right]$ % decrease in C.T. $= 1 - \sqrt{\dfrac{(1-D/P)}{1-\{D/P(1+M/100)\}}}$
Increased holding cost	Decrease in the cycle time = $\sqrt{\dfrac{2C(K+I)}{KID(1-D/P)}}\left[1-\sqrt{\dfrac{K+I(1+M/100)}{(1+M/100)(K+I)}}\right]$ % decrease in C.T. $= 1 - \sqrt{\dfrac{K+I(1+M/100)}{(1+M/100)(K+I)}}$
Decreased demand	Decrease in the cycle time = $\sqrt{\dfrac{2C(K+I)}{KID(1-D/P)}}\left[1-\sqrt{\dfrac{(1-D/P)}{(1-M/100)\{1-D(1-M/100)/P\}}}\right]$ % decrease in C.T. = $1 - \sqrt{\dfrac{(1-D/P)}{(1-M/100)\{1-D(1-M/100)/P\}}}$
Increased shortage cost	Decrease in the cycle time = $\sqrt{\dfrac{2C(K+I)}{KID(1-D/P)}}\left[1-\sqrt{\dfrac{I+K(1+M/100)}{(1+M/100)(K+I)}}\right]$ % decrease in C.T. $= 1 - \sqrt{\dfrac{I+K(1+M/100)}{(1+M/100)(K+I)}}$

A.3.2   Upward Variation in Cycle Time

Optimal cycle time increases because of:

    i. Increased facility setup cost
    ii. Reduced production rate
    iii. Reduced holding cost
    iv. Increased demand
    v. Reduced shortage cost

Now:

   M = % variation in parameter under consideration
   T_1 = Increased cycle time

    i. Increased setup cost:

Since the parameter under consideration is setup cost, M refers to the % increase in setup cost.

Now:

$$C_1 = \left(1 + \frac{M}{100}\right)C$$

Increase in the cycle time $= T_1 - T$

$$= \sqrt{\frac{2C_1(K+I)}{KID(1-D/P)}} - \sqrt{\frac{2C(K+I)}{KID(1-D/P)}}$$

$$= \sqrt{\frac{2C(K+I)}{KID(1-D/P)}} \left[\sqrt{\frac{C_1}{C}} - 1\right]$$

$$= \sqrt{\frac{2C(K+I)}{KID(1-D/P)}} \left[\sqrt{\left(1 + \frac{M}{100}\right)} - 1\right]$$

And:

$$\% \text{ increase in C.T.} = \sqrt{\left(1 + \frac{M}{100}\right)} - 1$$

In order to illustrate, consider the parameters of Example A.2 as follows:

D	C	I	P	K	T	E
600	50	35	960	100	0.131	763.76

After implementation of the increased setup cost, the effects are given as follows:

% Increase in C	5%	10%	15%	20%	25%	30%
C	52.50	55.00	57.50	60.00	62.50	65.00
T	0.134	0.137	0.140	0.143	0.146	0.149
% Increase in T	2.47%	4.88%	7.24%	9.54%	11.80%	14.02%
E	782.62	801.04	819.04	836.66	853.91	870.82
% Increase in E	2.47%	4.88%	7.24%	9.54%	11.80%	14.02%

ii. Reduced production rate:

$$P_1 \left(1 - \frac{M}{100}\right)P$$

Increase in the cycle time $= T_1 - T$

$$= \sqrt{\frac{2C(K+I)}{KID(1-D/P_1)}} - \sqrt{\frac{2C(K+I)}{KID(1-D/P)}}$$

$$= \sqrt{\frac{2C(K+I)}{KID(1-D/P)}} \left[\sqrt{\frac{(1-D/P)}{(1-D/P_1)}} - 1\right]$$

$$= \sqrt{\frac{2C(K+I)}{KID(1-D/P)}} \left[\sqrt{\frac{(1-D/P)}{1-\{D/P(1-M/100)\}}} - 1\right]$$

And:

$$\% \text{ increase in C.T.} = \sqrt{\frac{(1 - D/P)}{1 - \{D/P(1 - M/100)\}}} - 1$$

In order to illustrate, consider the reference parameters as follows:

D	C	I	P	K	T	E
600	50	35	960	100	0.131	763.76

After implementation of the reduced production rate, the effects in terms of approximate values are given as follows:

% Decrease in p	5%	10%	15%	20%	25%	30%
P	912	864	816	765	720	672
T	0.137	0.145	0.156	0.171	0.196	0.245
% Increase in T	4.70%	10.78%	19.02%	30.93%	50.00%	87.08%
E	729.50	689.43	641.69	583.33	509.18	408.25
% Decrease in E	4.49%	9.73%	15.98%	23.62%	33.33%	46.55%

iii. Reduced holding cost:

$$I_1 = \left(1 - \frac{M}{100}\right) I$$

Increase in the cycle time $= T_1 - T$

$$= \sqrt{\frac{2C(K + I_1)}{KI_1 D(1 - D/P)}} - \sqrt{\frac{2C(K + I)}{KID(1 - D/P)}}$$

$$= \sqrt{\frac{2C(K + I)}{KID(1 - D/P)}} \left[\sqrt{\frac{I(K + I_1)}{I_1(K + I)}} - 1\right]$$

$$= \sqrt{\frac{2C(K + I)}{KID(1 - D/P)}} \left[\sqrt{\frac{K + I(1 - M/100)}{(1 - M/100)(K + I)}} - 1\right]$$

and:

$$\% \text{ increase in C.T.} = \sqrt{\frac{K + I(1 - M/100)}{(1 - M/100)(K + I)}} - 1$$

In order to illustrate, consider the reference parameters as before:

D	C	I	P	K	T	E
600	50	35	960	100	0.131	763.76

After implementation of the reduced holding cost, the effects in terms of approximate values are provided below:

% Decrease in I	5%	10%	155	20%	25%	30%
I	33.25	31.50	29.75	28.00	26.25	24.50
T	0.134	0.138	0.142	0.146	0.151	0.156
% Increase in T	2.60%	5.41%	8.47%	11.80%	15.47%	19.52%
E	744.42	724.57	704.15	683.13	661.44	639.01
% Decrease in E	2.53%	5.13%	7.80%	10.56%	13.40%	16.33%

iv. Increased demand:

$$D_1 = \left(1 + \frac{M}{100}\right)D$$

Increase in the cycle time $= T_1 - T$

$$= \sqrt{\frac{2C(K+I)}{KID_1(1 - D_1/P)}} - \sqrt{\frac{2C(K+I)}{KID(1 - D/P)}}$$

$$= \sqrt{\frac{2C(K+I)}{KID(1 - D/P)}} \left[\sqrt{\frac{D(1 - D/P)}{D_1(1 - D_1/P}} - 1\right]$$

$$= \sqrt{\frac{2C(K+I)}{KID(1 - D/P)}} \left[\sqrt{\frac{(1 - D/P)}{(1 + M/100)\{1 - D(1 + M/100)/P\}}} - 1\right]$$

and:

$$\% \text{ increase in C.T.} = \sqrt{\frac{(1 - D/P)}{(1 + M/100)\{1 - D(1 + M/100)/P\}}} - 1$$

The reference parameters are as follows:

D	C	I	P	K	T	E
600	50	35	960	100	0.131	763.76

When demand is increased in this example, the effects in terms of approximate values are given below:

% Increase in D	5%	10%	15%	20%	25%	30.00%
D	630.00	660.00	690.00	720.00	750.00	780.00
T	0.133	0.137	0.141	0.146	0.153	0.162
% Increase in T	1.93%	4.45%	7.68%	11.80%	17.11%	24.03%
E	749.31	731.25	709.31	683.13	652.19	615.77
% Decrease in E	1.89%	4.26%	7.13%	10.56%	14.61%	19.38%

v. Reduced shortage cost:

$$K_1 = \left(1 - \frac{M}{100}\right)K$$

Increase in the cycle time = $T_1 - T$

$$= \sqrt{\frac{2C(K_1+I)}{K_1 ID(1-D/P)}} - \sqrt{\frac{2C(K+I)}{KID(1-D/P)}}$$

$$= \sqrt{\frac{2C(K+I)}{KID(1-D/P)}} \left[\sqrt{\frac{K(K_1+I)}{K_1(K+I)}} - 1\right]$$

$$= \sqrt{\frac{2C(K+I)}{KID(1-D/P)}} \left[\sqrt{\frac{I+K(1-M/100)}{(1-M/100)(K+I)}} - 1\right]$$

and:

$$\% \text{ increase in C.T.} = \sqrt{\frac{I+K(1-M/100)}{(1-M/100)(K+I)}} - 1$$

With the reference parameters as follows:

D	C	I	P	K	T	E
600	50	35	960	100	0.131	763.76

The effects in terms of approximate values are given below:

% Decrease in K	5%	10%	15%	20%	25%	30%
K	95	90	85	85	75	70
T	0.1318	0.1328	0.1339	0.1351	0.1365	0.1380
% Increase in T	0.68%	1.43%	2.26%	3.19%	4.23%	5.41%
E	758.60	752.99	746.87	740.15	732.76	724.57
% Decrease in E	0.68%	1.41%	2.21%	3.09%	4.06%	5.13%

Table A.4 summarizes the generalized results.

TABLE A.4

Results for Increased Cycle Time with Shortages

Variation in Input Parameter	Increase in Optimal C.T.
Increased facility setup cost	Increase in the cycle time $= \sqrt{\frac{2C(K+I)}{KID(1-D/P)}}\left[\sqrt{\left(1 + \frac{M}{100}\right)} - 1\right]$ % Increase in $$\text{C.T.} = \sqrt{\left(1 + \frac{M}{100}\right)} - 1$$
Reduced production rate	Increase in the cycle time $= \sqrt{\frac{2C(K+I)}{KID(1-D/P)}}\left[\sqrt{\frac{(1-D/P)}{1-\{D/P(1-M/100)\}}} - 1\right]$ % increase in $\text{C.T.} = \sqrt{\frac{(1-D/P)}{1-\{D/P(1-M/100)\}}} - 1$
Reduced holding cost	Increase in the cycle time $= \sqrt{\frac{2C(K+I)}{KID(1-D/P)}}\left[\sqrt{\frac{K+I(1-M/100)}{(1-M/100)(K+I)}} - 1\right]$ % Increase in $\text{C.T.} = \sqrt{\frac{K+I(1-M/100)}{(1-M/100)(K+I)}} - 1$
Increased demand	Increase in the cycle time $=$ $\sqrt{\frac{2C(K+I)}{KID(1-D/P)}}\left[\sqrt{\frac{(1-D/P)}{(1+M/100)\{1-D(1+M/100)/P\}}} - 1\right]$ % Increase in C.T. $=$ $\sqrt{\frac{(1-D/P)}{(1+M/100)\{1-D(1+M/100)/P\}}} - 1$
Reduced shortage cost	Increase in the cycle time $= \sqrt{\frac{2C(K+I)}{KID(1-D/P)}}\left[\sqrt{\frac{I+K(1-M/100)}{(1-M/100)(K+I)}} - 1\right]$ % Increase in C.T. $= \sqrt{\frac{I+K(1-M/100)}{(1-M/100)(K+I)}} - 1$

A.4   INTERACTION OF PARAMETERS WITHOUT SHORTAGES

When shortages are not allowed, production cycle time is determined on the basis of optimization related to the remaining various parameters. However, when a parameter varies, the cycle time also changes. In many situations either because of convenience in material handling or certain arrangement between buyer and supplier, a similar cycle time is preferred. Suitable management response should be available by way of change in another parameter.

A.4.1   INCREASED CYCLE TIME

Cycle time increases because of:

   i. Higher setup cost
  ii. Lower production rate
 iii. Lower holding cost

iv. Higher demand

i. Higher setup cost:

Consider the following parameters:
Annual demand, D = 600
Annual production holding cost per unit, I = ₹ 35
Annual production rate, P = 960
Setup cost, C = ₹ 50
From (A.4), optimal cycle time is:

$$T = \sqrt{\frac{2C}{DI(1 - D/P)}}$$

or

$$₹\ T = \sqrt{\frac{2 \times 50}{600 \times 35(1 - 600/960)}}$$

or

T = 0.1127 year
Now, if setup cost is increased by 10% then:
$C_1 = ₹55$

The corresponding cycle time will increase to 0.1182 year. However, if an objective is to have a similar cycle time as before, then a response might be in the form of production rate increase. The increased production rate can be obtained as follows:

$$\sqrt{\frac{2 \times 55}{600 \times 35(1 - 600/P_1)}} = 0.1127$$

or

$P_1 = 1021.11$
That is, approximately 6.37% increase in production rate.
For a general approach:
M = % variation in the parameter that triggers the change
N = % variation in the response parameter in order to have similar cycle time
Now:

$$C_1 = C\left(1 + \frac{M}{100}\right)$$

$$P_1 = P\left(1 + \frac{N}{100}\right)$$

For a similar cycle time:

$$\sqrt{\frac{2C}{DI(1 - D/P)}} = \sqrt{\frac{2C_1}{DI(1 - D/P)}}$$

or

$$C(1 - D/P_1) = C_1(1 - D/P)$$

or

$$1 - \frac{D}{P_1} = (1 + M/100)(1 - D/P)$$

or

$$\frac{D}{P_1} = 1 - (1 + M/100)(1 - D/P)$$

or

$$P_1 = \frac{D}{1 - (1 + M/100)(1 - D/P)}$$

or

$$1 + \frac{N}{100} = \frac{(D/P)}{1 - (1 + M/100)(1 - D/P)}$$

or

$$\frac{N}{100} = \frac{(D/P) - 1 + (1 + M/100)(1 - D/P)}{1 - (1 + M/100)(1 - D/P)}$$

or

$$\frac{N}{100} = \frac{(D/P) - 1 + 1 - (D/P) + (M/100) - (M/100)(D/P)}{1 - [(1 - (D/P)) + (M/100) - (M/100)(D/P)]}$$

or

$$\frac{N}{100} = \frac{(M/100)(1 - D/P)}{(D/P) - (M/100) + (M/100)(D/P)}$$

or

$$N = \frac{M(1 - D/P)}{(D/P)(1 + M/100) - (M/100)}$$

For the stated reference data, variation of N with respect to M is shown as follows:

S. No.	M	$N = \frac{M(1-D/P)}{(D/P)(1 + M/100) - (M/100)}$
1	5	3.09
2	10	6.38
3	15	9.89
4	20	13.64
5	25	17.65

The values of N are lower than that of M. However, these are more sensitive towards higher values of M.

ii. Lower production rate:

Reduction in the setup cost might be an option for a similar cycle time.

$$P_1 = P\left(1 - \frac{M}{100}\right)$$

$$C_1 = C\left(1 - \frac{N}{100}\right)$$

Now:

$$\sqrt{\frac{2C}{DI(1 - D/P)}} = \sqrt{\frac{2C_1}{DI(1 - D/P_1)}}$$

or

$$C_1(1 - D/P) = C(1 - D/P_1)$$

or

$$1 - \frac{N}{100} = \frac{(1 - D/P_1)}{(1 - D/P)}$$

or

$$\frac{N}{100} = \frac{(1 - D/P) - (1 - D/P_1)}{(1 - D/P)}$$

or

$$\frac{N}{100} = \frac{(D/P_1) - (D/P)}{(1 - D/P)}$$

or

$$\frac{N}{100} = \frac{D - D(P_1/P)}{P_1(1 - D/P)}$$

or

$$\frac{N}{100} = \frac{D[1 - 1 + (M/100)]}{P(1 - M/100)(1 - D/P)}$$

or

$$\frac{N}{100} = \frac{(D/P)(M/100)}{(1 - M/100)(1 - D/P)}$$

or

$$N = \frac{M(D/P)}{(1 - M/100)(1 - D/P)}$$

With the reference parameters as before:

D	C	I	P	T	E
600	50	35	960	0.113	887.41

And with the use of relevant parameters, the variation of N is shown below:

S. No.	M	$N = \dfrac{M(D/P)}{(1 - M/100)(1 - D/P)}$
1	2	3.40
2	4	6.94
3	6	10.64
4	8	14.49
5	10	18.52

The values of N are higher than that of M, and also are more sensitive towards higher values of M.

iii. Lower holding cost:

Increased production rate might be one of the options for a similar cycle time.

$$I_1 = I\left(1 - \frac{M}{100}\right)$$

$$P_1 = P\left(1 + \frac{N}{100}\right)$$

$$\sqrt{\frac{2C}{DI(1 - D/P)}} = \sqrt{\frac{2C}{DI_1(1 - D/P_1)}}$$

or

$$I_1(1 - D/P_1) = I(1 - D/P)$$

or

$$1 - \frac{D}{P_1} = \frac{(1 - D/P)}{(1 - M/100)}$$

or

$$\frac{D}{P_1} = \frac{1 - (M/100) - 1 + (D/P)}{(1 - M/100)}$$

or

$$P_1 = \frac{D(1 - M/100)}{(D/P) - (M/100)}$$

or

$$1 + \frac{N}{100} = \frac{(D/P)(1 - M/100)}{(D/P) - (M/100)}$$

or

$$\frac{N}{100} = \frac{(D/P)(1 - M/100) - (D/P) + (M/100)}{(D/P) - (M/100)}$$

or

$$\frac{N}{100} = \frac{(M/100) - (D/P)(M/100)}{(D/P) - (M/100)}$$

or

$$\frac{N}{100} = \frac{(M/100)(1 - D/P)}{(D/P - M/100)}$$

or

$$N = \frac{M(1 - D/P)}{(D/P) - (M/100)}$$

With the reference parameters including D and P as 600 and 960, respectively, the variation of N is provided below:

S. No.	M	$N = \frac{M(1 - D/P)}{(D/P) - (M/100)}$
1	2	1.24
2	4	2.56
3	6	3.98
4	8	5.50
5	10	7.14

The values of N are lower than that of M, but these are more sensitive towards higher values of M.

Another option can be a setup cost reduction:

$$I_1 = I\left(1 - \frac{M}{100}\right)$$

$$C_1 = C\left(1 - \frac{N}{100}\right)$$

$$\sqrt{\frac{2C}{DI(1 - D/P)}} - \sqrt{\frac{2C_1}{DI_1(1 - D/P)}}$$

or

$$C_1 I = C I_1$$

or

$$1 - \frac{N}{100} = 1 - \frac{M}{100}$$

or

$N = M$

iv. Higher demand:

Setup cost reduction and higher production rate are considered the options.

a. Setup cost reduction:

$$D_1 = D\left(1 + \frac{M}{100}\right)$$

$$C_1 = C\left(1 - \frac{N}{100}\right)$$

Now:

$$\sqrt{\frac{2C}{DI(1 - D/P)}} = \sqrt{\frac{2C_1}{D_1 I (1 - D_1/P)}}$$

or

$$C_1 D(1 - D/P) = CD_1(1 - D_1/P)$$

or

$$C_1 = \frac{C(1 + M/100)(1 - D_1/P)}{(1 - D/P)}$$

or

$$1 - \frac{N}{100} = \frac{(1 + M/100)(1 - D_1/P)}{(1 - D/P)}$$

or

$$\frac{N}{100} = \frac{(1 - D/P) - (1 + M/100)(1 - D_1/P)}{(1 - D/P)}$$

or

$$\frac{N}{100} = \frac{1 - (D/P) - (1 + M/100) + (1 + M/100)(D_1/P)}{(1 - D/P)}$$

or

$$\frac{N}{100} = \frac{(1 + M/100)^2(D/P) - (D/P) - (M/100)}{(1 - D/P)}$$

or

$$\frac{N}{100} = \frac{(D/P)[1 + (2M/100) + (M/100)^2 - 1] - (M/100)}{(1 - D/P)}$$

or

$$N = \frac{(D/P)[2M + (M^2/100)] - M}{(1 - D/P)}$$

With the reference parameters:

D	C	I	P	T	E
600	50	35	960	0.113	887.41

And with the use of relevant parameters, the variation of N is shown below:

S. No.	M	$N = \frac{(D/P)[2M+(M^2/100)]-M}{(1-D/P)}$
1	2	1.40
2	4	2.93
3	6	4.60
4	8	6.40
5	10	8.33

The values of N are lower than that of M, but these are more sensitive towards higher values of M.

(b) Higher production rate:

$$D_1 = D\left(1 + \frac{M}{100}\right)$$

$$P_1 = P\left(1 + \frac{N}{100}\right)$$

$$\sqrt{\frac{2C}{DI(1 - D/P)}} = \sqrt{\frac{2C}{D_1 I(1 - D_1/P)}}$$

or

$$D_1(1 - D_1/P_1) = D(1 - D/P)$$

or

$$1 - \frac{D_1}{P_1} = \frac{(1 - D/P)}{(1 + M/100)}$$

or

$$\frac{D_1}{P_1} = \frac{1 + (M/100) - 1 + (D/P)}{(1 + M/100)}$$

or

$$\frac{P_1}{D_1} = \frac{(1 + M/100)}{(M/100) + (D/P)}$$

or

$$P\left(1 + \frac{N}{100}\right) = \frac{D(1 + M/100)^2}{(M/100) + (D/P)}$$

or

$$1 + \frac{N}{100} = \frac{(D/P)(1 + M/100)^2}{(M/100) + (D/P)}$$

or

$$\frac{N}{100} = \frac{(D/P)(1 + M/100)^2 - (M/100) - (D/P)}{(M/100) + (D/P)}$$

or

$$\frac{N}{100} = \frac{(D/P)[1 + (2M/100) + (M/100)^2 - 1] - (M/100)}{(M/100) + (D/P)}$$

or

$$N = \frac{(D/P)[2M + (M^2/100)] - M}{(M/100) + (D/P)}$$

With the reference parameters as before, the variation of N is shown as follows:

S. No.	M	$N = \frac{(D/P)[2M+(M^2/100)]-M}{(M/100)+(D/P)}$
1	2	0.81
2	4	1.65
3	6	2.52
4	8	3.40
5	10	4.31

The values of N are much lower than that of M, but these are slightly more sensitive towards higher values of M. The N values are lower in comparison with the previous option. However, ease in implementation (that varies from organization to organization) should also be considered before finally selecting a suitable alternative.

A.4.2 DECREASED CYCLE TIME

Cycle time decreases because of:

  i. Lower setup cost
 ii. Higher production rate
iii. Higher holding cost
 iv. Lower demand

i. Lower setup cost:

Production rate decrease might be an option.

$$C_1 = C\left(1 - \frac{M}{100}\right)$$

$$P_1 = P\left(1 - \frac{N}{100}\right)$$

For a similar cycle time:

$$\sqrt{\frac{2C}{DI\,(1 - D/P)}} = \sqrt{\frac{2C_1}{DI\,(1 - D/P_1)}}$$

or

$$(1 - D/P_1)C = (1 - D/P)C_1$$

or

$$1 - \frac{D}{P_1} = (1 - D/P)\,(1 - M/100)$$

or

$$\frac{D}{P_1} = 1 - (1 - D/P)(1 - M/100)$$

or

$$P_1 = \frac{D}{1 - (1 - D/P)(1 - M/100)}$$

or

$$1 - \frac{N}{100} = \frac{(D/P)}{1 - (1 - D/P)(1 - M/100)}$$

or

$$\frac{N}{100} = \frac{1 - (1 - D/P)(1 - M/100) - (D/P)}{1 - (1 - D/P)(1 - M/100)}$$

or

$$\frac{N}{100} = \frac{(1 - D/P)(1 - 1 - M/100)}{1 - (1 - D/P)(1 - M/100)}$$

or

$$N = \frac{M(1 - D/P)}{1 - (1 - D/P)(1 - M/100)}$$

For the reference data:

D	C	I	P	T	E
600	50	35	960	0.113	887.41

Variation of N with respect to M is shown as follows:

S. No.	M	$N = \dfrac{M(1 - D/P)}{1 - (1 - D/P)(1 - M/100)}$
1	5	2.91
2	10	5.66
3	15	8.26
4	20	10.71
5	25	13.04

The values of N are lower than that of M, and are also less sensitive towards higher values of M.

ii. Higher production rate:

Holding cost reduction might be an option.

$$P_1 = P\left(1 + \frac{M}{100}\right)$$

$$I_1 = I\left(1 - \frac{N}{100}\right)$$

Now:

$$\sqrt{\frac{2C}{DI(1 - D/P)}} = \sqrt{\frac{2C}{DI_1(1 - D/P_1)}}$$

or

$$I_1(1 - D/P_1) = I(1 - D/P)$$

or

$$1 - \frac{N}{100} = \frac{(1 - D/P)}{(1 - D/P_1)}$$

or

$$\frac{N}{100} = \frac{(1 - D/P_1) - (1 - D/P)}{(1 - D/P_1)}$$

or

$$\frac{N}{100} = \frac{(D/P) - (D/P_1)}{(1 - D/P_1)}$$

or

$$\frac{N}{100} = \frac{D(P_1/P) - D}{(P_1 - D)}$$

or

$$\frac{N}{100} = \frac{D[(1 + M/100) - 1)]}{P(1 + M/100) - D}$$

or

$$\frac{N}{100} = \frac{(DM/100)}{(PM/100) + (P - D)}$$

or

$$N = \frac{DM}{(PM/100) + (P - D)}$$

With the reference parameters including D and P as 600 and 960, respectively, the variation of N is provided below:

S. No.	M	$N = \frac{DM}{(PM/100) + (P-D)}$
1	2	3.16
2	4	6.02
3	6	8.52
4	8	10.99
5	10	13.16

The values of N are higher than that of M, but these are less sensitive towards greater values of M.

iii. Higher holding cost:

Consider a production rate decrease for the management response.

$$I_1 = I\left(1 + \frac{M}{100}\right)$$

$$P_1 = P\left(1 - \frac{N}{100}\right)$$

$$\sqrt{\frac{2C}{DI(1 - D/P)}} = \sqrt{\frac{2C}{DI_1(1 - D/P_1)}}$$

or

$$I_1(1 - D/P_1) = I(1 - D/p)$$

or

$$1 - \frac{D}{P_1} = \frac{(1 - D/P)}{(1 + M/100)}$$

or

$$\frac{D}{P_1} = \frac{(1 + M/100) - 1 + (D/P)}{(1 + M/100)}$$

or

$$\frac{P1}{D} = \frac{(1 + M/100)}{(M/100) + (D/P)}$$

or

$$1 - \frac{N}{100} = \frac{(D/P)(1 + M/100)}{(M/100) + (D/P)}$$

or

$$\frac{N}{100} = \frac{(M/100) + (D/P) - (D/P) - (D/P)(M/100)}{(M/100) + (D/P)}$$

or

$$\frac{N}{100} = \frac{(M/100)(1 - D/P)}{(M/100) + (D/P)}$$

or

$$N = \frac{M(1 - D/P)}{(M/100) + (D/P)}$$

For the reference data:

D	C	I	P	T	E
600	50	35	960	0.113	887.41

Variation of N with respect to M is shown as follows:

S. No.	M	$N = \frac{M(1 - D/P)}{(M/100) + (D/P)}$
1	5	2.78
2	10	5.17
3	15	7.26
4	20	9.09
5	25	10.71

The values of N are lower than that of M, and are also less sensitive towards higher values of M.

iv. Lower demand:

Consider a reduction in holding cost for similar cycle time:

$$D_1 = D\left(1 - \frac{M}{100}\right)$$

$$I_1 = I\left(1 - \frac{N}{100}\right)$$

$$\sqrt{\frac{2C}{DI(1 - D/P)}} = \sqrt{\frac{2C}{D_1 I_1 (1 - D_1/P)}}$$

or

$$D_1 I_1 (1 - D_1/P) = DI (1 - D/P)$$

or

$$(1 - M/100)(1 - N/100) = \frac{(1 - D/P)}{(1 - D_1/P)}$$

or

$$1 - \frac{N}{100} = \frac{(1 - D/P)}{(1 - M/100)(1 - D_1/P)}$$

or

$$\frac{N}{100} = \frac{(1 - M/100)(1 - D_1/P) - (1 - D/P)}{(1 - M/100)(1 - D_1/P)}$$

or

$$\frac{N}{100} = \frac{1 - (D_1/P) - (M/100) + (M/100)(D_1/P) - 1 + (D/P)}{(1 - M/100)\{(1 - (D_1/P)(1 - M/100)\}}$$

or

$$\frac{N}{100} = \frac{(D/P) - (D_1/P) - (M/100)(1 - D_1/P)}{(1 - M/100)\{(1 - D/P)(1 - M/100)\}}$$

or

$$\frac{N}{100} = \frac{(D/P) - (D/P)(1 - M/100) - (M/100)(1 - D_1/P)}{(1 - M/100)\{1 - (D/P)(1 - M/100)\}}$$

or

$$\frac{N}{100} = \frac{(D/P)(1 - 1 + M/100) - (M/100)\{1 - (D/P)(1 - M/100)\}}{(1 - M/100)\{1 - (D/P)(1 - M/100)\}}$$

or

$$\frac{N}{100} = \frac{(D/P)(M/100) - (M/100)\{1 - (D/P)(1 - M/100)\}}{(1 - M/100)\{1 - (D/P)(1 - M/100)\}}$$

or

$$N = \frac{M\,(D/P) - M\,\{1 - (D/P)(1 - M/100)\}}{(1 - M/100)\{1 - (D/P)(1 - M/100)\}}$$

With the reference parameters including D and P as 600 and 960, respectively, the variation of N is provided below:

S. No.	M	$N = \frac{M\,(D/P) - M\,\{1 - (D/P)(1 - M/100)\}}{(1 - M/100)\{1 - (D/P)(1 - M/100)\}}$
1	2	1.25
2	4	2.34
3	6	3.29
4	8	4.09
5	10	4.76

The values of N are lower than that of M, and also are less sensitive towards greater values of M.

A.5   INTERACTION OF PARAMETERS WITH SHORTAGES

With the inclusion of shortages, the interaction is analyzed concerning various parameters.

A.5.1   LONGER CYCLE TIME

Cycle time increases also because of a reduction in shortage cost. For a similar cycle time, consider the management response in the form of:

  i. Reduced setup cost
  ii. Increased production rate

For a general approach, let:
  M = % variation in the shortage cost
  N = % variation in the response parameter

  i. Reduced setup cost:

$$K_1 = K\left(1 - \frac{M}{100}\right)$$

$$C_1 = C\left(1 - \frac{N}{100}\right)$$

Since the optimal cycle time is given as:

$$T = \sqrt{\frac{2C\,(K + I)}{KID\,(1 - D/P)}}$$

For a similar cycle time:

$$\sqrt{\frac{2C\,(K + I)}{KID\,(1 - D/P)}} = \sqrt{\frac{2C_1\,(K_1 + I)}{K_I\,ID\,(1 - D/P)}}$$

or

$$\frac{C\,(K + I)}{K} = \frac{C_1\,(K_1 + I)}{K_1}$$

or

$$c_1 = \frac{K_1 C\,(K + I)}{K\,(K_1 + I)}$$

or

$$1 - \frac{N}{100} = \frac{(1 - M/100)(K + I)}{I + K\,(1 - M/100)}$$

or

$$\frac{N}{100} = \frac{I + K\,(1 - M/100) - (1 - M/100)(K + I)}{I + K\,(1 - M/100)}$$

or

$$\frac{N}{100} = \frac{I - I\,(1 - M/100)}{I + K\,(1 - M/100)}$$

or

$$\frac{N}{100} = \frac{(IM/100)}{I + K\,(1 - M/100)}$$

or

$$N = \frac{IM}{I + K\,(1 - M/100)}$$

For the reference parameters as follows:

D	C	I	P	K	T	E
600	50	35	960	100	0.131	763.76

The values of N are given w.r.t. M as follows:

S. No.	M	$N = \dfrac{IM}{I + K(1 - M/100)}$
1	5	1.35
2	10	2.80
3	15	4.38
4	20	6.09
5	25	7.95

The values of N are much lower than that of M, but are more sensitive towards greater values of M.

$$K_1 = K\left(1 - \frac{M}{100}\right)$$

$$P_1 = P\left(1 + \frac{N}{100}\right)$$

ii. Increased production rate:

For a similar cycle time:

$$\sqrt{\frac{2C(K + I)}{KID(1 - D/P)}} = \sqrt{\frac{2C(K_1 + I)}{K_1 \, ID(1 - D/P_1)}}$$

or

$$\frac{(K + I)}{K(1 - D/P)} = \frac{(K_1 + I)}{K_1(1 - D/P_1)}$$

With the use of the above expression and following the process explained before, it can be analyzed furthermore. However, in certain cases, a combination of parameters may also be useful. For example, with reference to the previous option (i.e., the reduced setup cost):

M = 15; N = 4.38

However, in certain operational settings, if the management feels that the value of N cannot be more then three (say), then a combination of parameters may be considered.

Combination of setup cost and production rate:

Let M, N, and R be the % variation in shortage cost, setup cost, and production rate.

Now:

$$K_1 = K\left(1 - \frac{M}{100}\right)$$

$$C_1 = C\left(1 - \frac{N}{100}\right)$$

$$P_1 = P\left(1 + \frac{R}{100}\right)$$

For a similar cycle time:

$$\sqrt{\frac{2C(K+I)}{KID(1 - D/P)}} = \sqrt{\frac{2C_1(K_1+I)}{K_I\,ID(1 - D/P_1)}}$$

or

$$\frac{C(K+I)}{K(1 - D/P)} = \frac{C_1(K_1+I)}{K_1(1 - D/P_1)}$$

or

$$1 - \frac{D}{P_1} = \frac{KC_1(K_1+I)(1 - D/P)}{K_1C(K+I)}$$

or

$$\frac{D}{P_1} = \frac{K_1C(K+I) - KC_1(K_1+I)(1 - D/P)}{K_1C(K+I)}$$

or

$$\frac{P_1}{D} = \frac{K_1C(K+I)}{K_1C(K+I) - KC_1(K_1+I)(1 - D/P)}$$

or

$$1 + \frac{R}{100} = \frac{(D/P)K_1C(K + I)}{K_1C(K + I) - KC_1(K_1 + I)(1 - D/P)}$$

or

$$\frac{R}{100} = \frac{(D/P)K_1C(K + I) - K_1C(K + I) + KC_1(K_1 + I)(1 - D/P)}{K_1C(K + I) - KC_1(K_1 + I)(1 - D/P)}$$

or

$$\frac{R}{100} = \frac{KC_1(K_1 + I)(1 - D/P) - K_1C(K + I)(1 - D/P)}{K_1C(K + I) - KC_1(K_1 + I)(1 - D/P)}$$

or

$$\frac{R}{100} = \frac{(1 - D/P)[KC_1(K_1 + I) - K_1C(K + I)]}{KC[(1 - M/100)(K + I) - (1 - N/100)(K_1 + I)(1 - D/P)]}$$

or

$$\frac{R}{100} = \frac{(1 - D/P)KC[(1 - N/100)(K_1 + I) - (1 - M/100)(K + I)]}{KC[(1 - M/100)(K + I) - (1 - N/100)(1 - D/P)\{K(1 - M/100) + I\}]}$$

or

$$R = \frac{100(1 - D/P)[(1 - N/100)\{K(1 - M/100) + I\} - (1 - M/100)(K + I)]}{[(1 - M/100)(K + I) - (1 - N/100)(1 - D/P)\{K(1 - M/100) + I\}]}$$

With the reference parameters:

D	C	I	P	K	T	E
600	50	35	960	100	0.131	763.76

For example, if M = 15, then:

$$R = \frac{4.375 - N}{1.55 + 0.01N}$$

For M = 15, the various combinations of R and N are given as follows:

S. No.	N	$R = \frac{4.375 - N}{1.55 + 0.01N}$
1	1	2.16
2	2	1.51
3	3	0.87
4	4	0.24

A.5.2 SHORTER CYCLE TIME

Cycle time decreases also because of an increase in shortage cost. For a similar cycle time, consider the management response in the form of:

i. Reduction in the holding cost
ii. Reduction in the production rate

It can be analyzed independently. However, as discussed before, a combination of parameters may also be useful.

Combination of holding cost and production rate:

Let M, N, and R be the % variation in shortage cost, holding cost, and production rate.

Now:

$$K_1 = K\left(1 + \frac{M}{100}\right)$$

$$I_1 = I\left(1 - \frac{N}{100}\right)$$

$$P_1 = P\left(1 - \frac{R}{100}\right)$$

For a similar cycle time:

$$\sqrt{\frac{2C(K + I)}{KID(1 - D/P)}} = \sqrt{\frac{2C(K_1 + I)}{K_1 I_1 D(1 - D/P_1)}}$$

or

$$\frac{(K + I)}{KI\,(1 - D/P)} = \frac{(K_1 + I_1)}{K_1 I_1 (1 - D/P_1)}$$

or

$$1 - \frac{D}{P_1} = \frac{KI\,(K_1 + I_1)(1 - D/P)}{K_1 I_1 (K + I)}$$

or

$$\frac{D}{P_1} = \frac{K_1 I_1 (K + I) - KI\,(K_1 + I_1)(1 - D/P)}{K_1 I_1 (K + I)}$$

or

$$\frac{P_1}{D} = \frac{K_1 I_1 (K + I)}{K_1 I_1 (K + I) - K\,I\,(K_1 + I_1)(1 - D/P)}$$

or

$$1 - \frac{R}{100} = \frac{(D/P)K_1 I_1 (K + I)}{K_1 I_1 (K + I) - KI\,(K_1 + I_1)(1 - D/P)}$$

or

$$\frac{R}{100} = \frac{K_1 I_1 (K + I) - KI\,(K_1 + I_1)(1 - D/P) - (D/P)K_1 I_1 (K + I)}{K_1 I_1 (K + I) - KI\,(K_1 + I_1)(1 - D/P)}$$

or

$$\frac{R}{100} = \frac{K_1 I_1 (K + I)(1 - D/P) - KI\,(K_1 + I_1)(1 - D/P)}{K_1 I_1 (K + I) - KI\,(K_1 + I_1)(1 - D/P)}$$

or

$$\frac{R}{100} = \frac{(1 - D/P)[K_1 I_1 (K + I) - KI\,(K_1 + I_1)]}{K_1 I_1 (K + I) - KI\,(K_1 + I_1)(1 - D/P)}$$

or

$$\frac{R}{100} = \frac{(1 - D/P)[(1 + M/100)(1 - N/100)(K + I) - (K_1 + I_1)]}{(1 + M/100)(1 - N/100)(K + I) - (K_1 + I_1)(1 - D/P)}$$

or

$$R = \frac{100(1 - D/P)[(1 + M/100)(1 - N/100)(K + I) - \{K(1 + M/100) + I(1 - N/100)\}]}{(1 + M/100)(1 - N/100)(K + I) - [(1 - D/P)\{K(1 + M/100) + I(1 - N/100)\}]}$$

With the reference parameters:

D	C	I	P	K	T	E
600	50	35	960	100	0.131	763.76

For example, if $M = 15$, then:

$$R = \frac{65.625 - 15.03125N}{33 - 0.47375N}$$

For $M = 15$, the various combinations of N and R are given as follows:

S. No.	N	$R = \frac{65.625 - 15.03125N}{33 - 0.47375N}$
1	1	1.55
2	2	1.11
3	3	0.65
4	4	0.18

General Reading

Inventory planning is done at the product level as well as at the component level. Certain aspects are discussed now in this context. On the shop floor, there are jobs available for processing and a need exists to sequence them suitably.

1 SEQUENCING OF JOBS ON A SINGLE MACHINE

Sequencing of jobs on a single machine is also called as single machine scheduling. Various jobs are ready to be processed on a machine with individual due dates of completion. Following are the parameters with respect to each job:

i. Ready time (R_j): This is the time when a job is ready for processing on a machine, in other words, when it is released for processing.

ii. Due time (D_j): This is the due time or due date for completion of any job J. If the job is completed after the due date, it gets delayed, i.e., it is a tardy job.

iii. Processing time (t_j): This is the duration a job will take on a machine, i.e., the operation time.

From these input parameters, certain data are derived. The derived data are defined, considering the following example.

Three jobs, 1, 2, and 3, are to be processed on a machine in the same sequence. All the jobs are ready now, i.e., at time zero $(R_j = 0)$. If due dates are provided, then the remaining time from now (i.e., at time zero) can easily by computed. Due time (D_j) is directly given as follows, along with the processing time (t_j).

Job	D_j	t_j
1	2	1
2	4	6
3	5	2

i. Completion time (C_j): This is the cumulative value, i.e., the cumulative sum of t_j.

J	t_j	C_j
1	1	1
2	6	7
3	2	9

Completion time for the first job in the sequence is similar to processing time; for the second job, i.e., $C_2 = C_1 + t_2 = 1 + 6 = 7$. Similarly, $C_3 = C_2 + t_3 = 7 + 2 = 9$. Therefore:

$$C_j = C_{j-1} + t_j$$

Average completion time:

$$\bar{C} = \frac{\Sigma C_j}{n}$$

where n = number of jobs. Thus:

$$\bar{C} = \frac{1 + 7 + 9}{3} = 5.67$$

ii. Flow time (F_j): Flow time is the difference between completion time of a job, C_j, and ready time of that job, R_j.

$$F_j = C_j - R_j$$

This is the time duration in which the job flows on the shop floor. As $R_j = 0$, the flow time is similar to completion time, i.e., $F_j = C_j$.
   Average or mean flow time:

$$\bar{F} = \frac{\Sigma F_j}{n}$$

Maximum flow time, $F_{max} = 9$.

iii. Queue time (Q_j): This is the duration in which the job remains in queue before actual processing:

$$Q_j = F_j - t_j$$

J	t_j	F_j	Q_j
1	1	1	0
2	6	7	1
3	2	9	7

Average queue time:

$$\bar{Q} = \frac{\Sigma Q_j}{n} = \frac{0 + 1 + 7}{3} = 2.67$$

Maximum queue time, Q_{max} is 7, corresponding to job 3.

iv. Number of jobs in process during certain time (P_n): This relates to work in process. Three jobs have been released to the shop floor before a machine M.

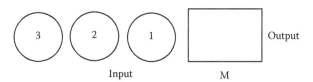

Now job 1 is processed on machine M for 1 unit of time as $t_j = 1$.

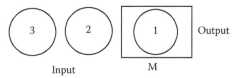

Three jobs are in process for time 1. Now, processing of job 1 is over and job 2 will be processed on the machine for 6 units of time.

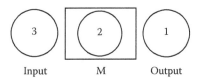

Two jobs (viz., 2 and 3) are in process for 6 units of time. Finally, one job, i.e., job 3, will be processed for 2 units of time ($t_3 = 2$).

Jobs are considered to be in process when either they await processing or are under actual operation over the machine.

Now:

3 jobs are in process for 1 unit of time as $t_1 = 1$

2 jobs are in process for 6 units of time as $t_2 = 6$

1 job is in process for 2 units of time as $t_3 = 2$

Jobs will continue processing for the maximum flow time, i.e., $F_{max} = 9$.

Therefore, the mean number of jobs in process:

$$\bar{P} = \frac{(3 \times 1) + (2 \times 6) + (1 \times 2)}{9} = 1.89$$

That is:

$$P = \frac{\Sigma(\text{Number of jobs X time for which they are in process})}{\Gamma_{max}}$$

If there are n jobs:

$$P = \frac{[nXt_1] + [(n-1)Xt_2] + [(n-2)Xt_3] + \dots + [2Xt_{n-1}] + [1Xt_n]}{F_{max}}$$

v. Lateness (L_j): This is a measure by which it can be known whether a job is delayed or it is completed before the due time.

$L_j = C_j - D_j$

J	D_j	C_j	$L_j = C_j - D_j$
1	2	1	-1
2	4	7	3
3	5	9	4

For the first job, lateness is negative, i.e., −1, as this job is completed before the due time.

Maximum lateness, L_{max}, is 4, corresponding to job 3.

Average lateness:

$$\bar{L} = \frac{-1 + 3 + 4}{3} = 2$$

vi. Tardiness (T_j): A job will be considered tardy if it is delayed, i.e., it is completed after the due time. The job is not tardy if it is completed on or before the due date. If lateness is positive, then the tardiness T_j is similar to L_j (otherwise T_j is zero).

J	L_j	T_j
1	−1	0
2	3	3
3	4	4

For L_j being negative or zero, tardiness T_j is zero (otherwise $T_j = L_j$).

Considering the column T_j (values 0, 3, 4), maximum tardiness T_{max} is 4, corresponding to job 3.

Different measures of performance are as follows:

Average completion time – $\bar{C}$

Average flow time – $\bar{F}$

Average queue time – $\bar{Q}$
Maximum queue time – Q_{max}
Mean number of jobs in process – $\bar{P}$
Average lateness – $\bar{L}$
Maximum lateness – L_{max}
Maximum tardiness – T_{max}

1.1 SPT RULE

The shortest processing time (SPT) rule suggests that the job with the shortest processing time should be sequenced first. In other words, the jobs are sequenced for operations on a machine in the order of increasing processing time. Jobs are arranged in the increasing order of processing time t_j.

The following measures of performance are minimized using the SPT rule, assuming that all the jobs are ready at the same time:

Average completion time – $\bar{C}$
Average flow time – $\bar{F}$
Average queue time – $\bar{Q}$
Maximum queue time – Q_{max}
Mean number of jobs in process – $\bar{P}$
Average lateness – $\bar{L}$

Example 1: The following seven jobs are ready at the same time, i.e., at time zero, for processing on a single machine. Apply the SPT rule for sequencing of jobs and evaluate different measures of performance.

J:	1	2	3	4	5	6	7
D_j	13	9	4	23	6	19	8
t_j	8	3	2	11	4	10	6

Now to apply the SPT rule, jobs are arranged in the increasing order of processing time t_j.

J:	3	2	5	7	1	6	4
D_j	4	9	6	8	13	19	23
t_j	2	3	4	6	8	10	11
C_j	2	5	9	15	23	33	44
F_j	2	5	9	15	23	33	44
Q_j	0	2	5	9	15	23	33
L_j	-2	-4	3	7	10	14	21

Measures of performance which will be minimized are evaluated as follows:

$$\bar{C} = \frac{\Sigma C_j}{n} = \frac{131}{7} = 18.71$$

$$\bar{F} = \frac{\Sigma F_j}{n} = \frac{131}{7} = 18.71$$

$$\bar{Q} = \frac{\Sigma Q_j}{n} = \frac{87}{7} = 12.43$$

$Q_{max} = 33$

The mean number of jobs in process:

$$\bar{P} = \frac{[7Xt_3] + [6Xt_2] + [5Xt_5] + [4Xt_7] + [3Xt_1] + [2Xt_6] + [1Xt_4]}{F_{max}}$$

$$= \frac{[7X2] + [6X3] + [5X4] + [4X6] + [3X8] + [2X10] + [1X11]}{44}$$

$$= \frac{131}{44} = 2.98$$

$$\bar{L} = \frac{\Sigma L_j}{n}$$

$$= \frac{-2 - 4 + 3 + 7 + 10 + 14 + 21}{7}$$

$$= \frac{49}{7} = 7$$

a. Average completion time:
b. Average flow time:
c. Average queue time:
d. Maximum queue time:
e. As $n = 7$ and as the sequence of jobs is $3 - 2 - 5 - 7 - 1 - 6 - 4$.
f. Average lateness:

1.2   EDD ORDER

Jobs are sequenced in the order of their earliest due date (EDD) first. Jobs are arranged in the increasing order of due date or due time, D_j.

The EDD order minimizes the following measures of performance:

   i. Maximum lateness – L_{max}
  ii. Maximum tardiness – T_{max}

Example 2: Consider the input data of the previous example and sequence the jobs in EDD order. Evaluate the performance measures.

Jobs are arranged in EDD order as follows.

J:	3	5	7	2	1	6	4
D_j	4	6	8	9	13	19	23
t_j	2	4	6	3	8	10	11
C_j	2	6	12	15	23	33	44
L_j	−2	0	4	6	10	14	21
T_j	0	0	4	6	10	14	21

$L_{max} = 21$ and $T_{max} = 21$, whereas average lateness, $\bar{L}$, is obtained as 7.57.

2   METHOD TO MINIMIZE THE NUMBER OF TARDY JOBS

If the objective is to minimize the number of tardy jobs, then the following method is adopted:

1. Imagine two locations: X and Y. X represents the set of jobs that are sequenced in EDD order and initially Y is an empty set of jobs.
2. If no tardy jobs are found in X, then this is the desired sequence. Otherwise, the first tardy job in X is identified and this is called the pth job.
3. Now consider only the first P jobs in set X and the job with maximum processing time is picked among the first P jobs. This is shifted to another set Y. But in this process, calculations previously made will be disrupted. Completion time, C_j, and tardiness, T_j, are recomputed.

The process from step (2) onwards is repeated until no tardy jobs are found in set X, i.e., all the jobs in set X will have tardiness T_j as zero.

Example 3: Consider the input data of Example 1 and apply the method to minimize the number of tardy jobs.

As the arrangement of jobs in EDD order is the essential requirement to apply the iterative method, computations will be made accordingly. This process is already completed in Example 2 and the computations are reproduced below. Sets X and Y are marked where Y is the empty set at present.

Steps (2) and (3) are to be repeated in each cycle until all jobs in set X have zero tardiness. Refer to Table 1. Only two jobs, 3 and 5, are non-tardy jobs out of 7 jobs.

TABLE 1
Computation Corresponding to EDD Order

			X					Y
J:	3	5	7	2	1	6	4	
D_j	4	6	8	9	13	19	23	
t_j	2	4	6	3	8	10	11	
C_j	2	6	12	15	23	33	44	
L_j	-2	0	4	6	10	14	21	
T_j	0	0	4	6	10	14	21	

TABLE 2
Computation at the End of Cycle 1

			X				Y
J:	3	5	2	1	6	4	7
D_j	4	6	9	13	19	23	8
t_j	2	4	3	8	10	11	6
C_j	2	6	9	17	27	38	44
L_j	-2	0	0	4	8	15	36
T_j	0	0	0	4	8	15	36

CYCLE 1

The sequence of jobs is $3 - 5 - 7 - 2 - 1 - 6 - 4$, and the first tardy job is job 7 with $T_j = 4$; job 7 is third in the sequence; thus, $p = 3$.

Now consider only the first three jobs in the sequence, i.e., jobs 3, 5, and 7, with processing time $t_j = 2$, 4, and 6, respectively. The maximum processing time is 6, corresponding to job 7; therefore, job 7 is shifted to set Y. Now the sequence is $3 - 5 - 2 - 1 - 6 - 4 - 7$. Recompute the parameters, as shown in Table 2.

CYCLE 2

Consider Table 2 for the next iteration. The first tardy job is 1, which is fourth in the sequence. The processing time of first four jobs are 2, 4, 3, and 8. The maximum processing time is 8, corresponding to job 1 and job 1 is shifted to set Y, as shown in Table 3.

TABLE 3
Computation for Cycle 2

		X					Y
J:	3	5	2	6	4	7	1
D_j	4	6	9	19	23	8	13
t_j	2	4	3	10	11	6	8
C_j	2	6	9	19	30	36	44
L_j	−2	0	0	0	7	28	31
T_j	0	0	0	0	7	28	31

TABLE 4
Computation for Cycle 3

		X					Y	
J:	3	5	2	6	7	1	4	
D_j	4	6	9	19	8	13	23	
t_j	2	4	3	10	6	8	11	
C_j	2	6	9	19	25	33	44	
L_j	−2	0	0	0	17	20	21	
T_j	0	0	0	0	17	20	21	

CYCLE 3

In Table 3, the first tardy job is 4, which is fifth in the sequence. In the first five jobs, a maximum processing time is 11, corresponding to job 4. Job 4 is shifted to Y, as shown in Table 4.

No tardy job is in set X; therefore, this is the desired sequence. The sequence is $3 - 5 - 2 - 6 - 7 - 1 - 4$, with three tardy jobs, 7, 1, and 4, with tardiness values 17, 20, and 21, respectively.

3   TWO-MACHINE FLOW SHOP SCHEDULING

As shown in Figure 1, all the jobs are processed on machines 1 and 2 sequentially.

A number of jobs are on the input side of the two-machine flow shop. The processing times of all jobs on both machines are known. The time when the last job will be completed in the flow shop is of interest. This is called make span, i.e., the maximum completion time, C_{max}. The objective is to determine the sequence of

Input

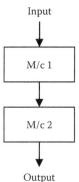

M/c 1

M/c 2

Output   **FIGURE 1**   Two-machine pure flow shop.

jobs in order to minimize the make span, and Johnson's algorithm is used for this objective. In this algorithm, the following steps are used:

1. Processing times of all the jobs on each machine are observed and the least processing time is selected. If this least processing time corresponds to machine 1, go to step 2. If this least processing time is on machine 2, go to step 3.
2. Jobs corresponding to the least processing time will occupy the earliest available space in the sequence. Go to step 4.
3. Jobs corresponding to the least processing time will occupy the latest available space in the sequence.
4. Keep the job, which has occupied the respective space in the sequence, is out of consideration for now. For the remaining jobs, repeat steps (1) through (4) until all jobs are accommodated in the sequence.

Example 4: A two-machine flow shop is interested in sequencing jobs so that the make span is minimized. The operation times of all six jobs on each machine are as follows.

Job	Time	
	Machine 1	Machine 2
J_1	8	5
J_2	6	7
J_3	4	8
J_4	15	14
J_5	8	9
J_6	11	10

As there are six jobs, six available spaces are there in the sequence, as shown below.

_ _ _ _ _ _

Cycle 1
The operation time on both machines is observed. The least time is 4 on machine 1, which corresponds to job J_3. This job occupies the earliest available space in the sequence.

J_3 _ _ _ _ _

As job J_3 is placed in the sequence, this is eliminated.

Cycle 2
The remaining jobs are J_1, J_2, J_4, J_5, and J_6. From their processing times, the least is 5, corresponding to job J_1. As this time is on machine 2, assign the job to the latest available space in the sequence:

J_3 _ _ _ _ J_1

Cycle 3
Out of the processing time of the remaining jobs, J_2, J_4, J_5, and J_6 on both machines, 6 is the least time on machine 1, corresponding to job J_2, and it is placed at the earliest available space:

J_3 J_2 _ _ _ J_1

Cycle 4
Consider the remaining jobs J_4, J_5, and J_6. The least time is 8 on machine 1 for job J_5, and assigned to the earliest available space:

J_3 J_2 J_5 _ _ J_1

Cycle 5
The remaining jobs are J_4 and J_6.Processing times are 15 and 14 on machines 1 and 2, respectively, for J_4. The processing times on machines 1 and 2 are 1 and 10, respectively for J_6. Out of the processing time (15, 14, 11, 10), 10 is the least, corresponding to machine 2 for job J_6. J_6 is assigned to the latest available space:

J_3 J_2 J_5 _ J_6 J_1

Cycle 6
Only one job, J_4, is left, which will take the last available position in the sequence:

J_3 J_2 J_5 J_4 J_6 J_1

To minimize the make span, the jobs are scheduled in the above order.

(Note: If there is a tie in selecting the least processing time, assign any one job arbitrarily.)

Now, to evaluate the make span, a Gantt chart is prepared for this schedule, as follows.

0	4	10	18	33	44	52
Machine 1	J_3 (4)	J_2 (6)	J_5 (8)	J_4 (15)	J_6 (11)	J_1 (8)

0	4	12	19	28	33	47	57	62
Machine 2	Idle	J_3 (8)	J_2 (7)	J_5 (9)	I-dle	J_4 (14)	J_6 (10)	J_1 (5)

Individual time taken by each job on machines 1 and 2 is given in brackets, along with jobs and cumulative values of time, are shown on the top for each machine in the Gantt chart.

The last job in the sequence is J_1, which will be completed at time 62. Therefore, the make span is 62 units of time.

Initially, machine 2 will be idle for 4 units of time as the first job, J_3, is being processed on machine 1. Machine 2 is also idle from time 28 to 33 as job J_4 is not ready for processing. Job J_4 is completed on machine 1 at time 33 and then it will start processing on machine 2.

Sequencing of jobs is also summarized as follows.

Cycle	Remaining Jobs to Be Scheduled	Least Processing Time	Sequence
1	J_1, J_2, J_3, J_4, J_5, J_6	4	J_3 _ _ _ _ _
2	J_1, J_2, J_4, J_5, J_6	5	J_3 _ _ _ _ J_1
3	J_2, J_4, J_5, J_6	6	J_3 J_2 _ _ _ J_1
4	J_4, J_5, J_6	8	J_3 J_2 J_5 _ _ J_1
5	J_4, J_6	10	J_3 J_2 J_5 _ J_6 J_1
6	J_4	14	J_3 J_2 J_5 J_4 J_6 J_1

In the context of internal logistics, including the material flow and processing on the shop floor, productivity aspects are relevant. Work study may also be a set of techniques used to raise the productivity of a company.

4   WORK STUDY

In general, productivity is defined as output/input. To increase the productivity, either output should be increased with the same level of input or input should be decreased for the same level of output. Productivity may also be increased by changing the output and input in such a way that the ratio of output to input is increased. For the profitability and improvement in the operations, continuous enhancement in the productivity becomes necessary. Work study is a set of techniques used to raise the productivity of an organization.

As the work study brings a change in the existing method of doing a task and normally human beings resist change, a suitable environment should be created before conducting the work study. Employees in the organization should be informed regarding the benefits of the work study, i.e., the productivity enhancement. By better utilization of resources, profitability will increase and with increased salaries and better growth prospects, it is for the welfare of employees. Work study is divided into two kinds of studies such as motion study and time study.

4.1 MOTION STUDY

Motion study may also be called method study. Its objective is to develop a more effective method and subsequent reduction in cost. This is achieved by systematic recording and critical examination of an existing as well as proposed method of performing a task. Steps to be followed for conducting a method study are discussed below:

a. The whole job is to be divided into elements or smaller tasks.
b. Each element of the industrial operation is to be examined critically. Some of the questions that may be asked are:
   i.   Why this material is being used for processing? Is there any possibility of substituting this material?
   ii. Why is an operation conducted in a particular manner only? Is there any other way of doing this task?
   iii. Is there ease in operation if it is conducted on some other equipment or machine that is available?
c. Critical examination of the operation will lead to a modification in the existing method. Try to modify the existing method or develop a new method.
d. A modified method or new proposed method needs training as well as arrangement of various facilities. Install the proposed method. Supervise the operations closely for a longer duration in order to maintain the proposed method. This is necessary because people usually revert to old practices.

The following five basic symbols are used in a motion study.

Description	Symbol
Operation	○
Inspection	□
Delay Storage	D ▽
Transport	⇨

Sometimes an operation is going on and, simultaneously, inspection may also be carried out. To represent this "operation and inspection", the following symbol is used:

4.2   CHARTS

There are certain charts which are useful in method study. These may be viewed as recording techniques. Following are various types of charts:

a. Operation chart: Operations and inspection to be carried out on any product are recorded as shown below:

Time   Symbol   Description

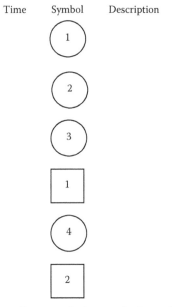

Suppose that a product requires four numbers of operations, and inspection at two stages. One, after operation 3 is over and another, after operation 4 is over, which is also the final inspection.

After operation 3, the first inspection takes place, and that is why it is numbered as 1 in the symbol of inspection. On the left side, time taken for operation/inspection is recorded. On the right side, a description of each operation/inspection is mentioned.

b. Flow process chart: This chart indicates how a product flows inside the industrial organization, but it may also be developed corresponding to activities of a person. A flow process chart is shown.

Time Taken	Distance Traveled	Description	○	□	D	▽	⇨
–	–	Item in storage					
3 min	200 m	Item transported to shop floor					
15 min	–	Processing of the item					
4 min	–	Inspecting the item					
2 min	–	Item awaits for further movement					
3 min	200 m	Transporting the item to storage					
–	–	Item in store					

Initially an item is in storage, and then it is transported to the shop floor for processing. Transportation time is 3 min and distance traveled is 200 m. The corresponding points below "storage" and "transport" symbols are joined by a straight line. Following a similar procedure, corresponding points are joined. The time taken in each activity as well as distance traveled, if any, are also recorded as shown, along with the description of each activity. If the item spends some time awaiting for any activity, it is recorded as a "delay".

- String diagram: A drawing of the layout using a suitable scale is prepared on the drawing sheet. A drawing thus prepared is placed on the wooden board and pins, which represent significant work centers, are fixed at appropriate places on the sheet. Movement of the operator is recorded using a thread or string. Suppose that the person moves from work center A to work center B. Then the thread is tied to the pin representing A, and afterwards to the pin representing B on the drawing sheet. In this way, the thread is tied as per the movement during a specified period. The objective is to know how much distance is traveled between any two facilities, which may be determined by measuring the length of the thread considering the scale of the drawn layout. A string diagram is useful when more persons are moving in a certain area or a person has to go to two or more numbers of facilities often. If the total distance traveled between the two facilities X and Y is too much, and these two facilities are far apart, then these may be brought closer to each other if feasible.
- Two-hand chart: This is also known as a left hand-right hand chart. The focus is on recording the functions of both hands. The operation under consideration is studied by observing the movement of hands and time is recorded using a clock. Clock readings are recorded on the central portion of the paper and a description of left-hand and right-hand motions are mentioned on left- and right-hand sides of the paper, respectively.

• Simo chart: A simultaneous motion (Simo) chart is similar to the two-hand chart except that the time needed for movement of the hands is recorded using a suitable time scale.

Micromotion Study

In the micromotion study, a permanent record in the form of a film is obtained using a camera with a clock. The film can be viewed at any time later whenever needed and the details of the operation can be observed.

Cyclegraph

A small bulb is attached to any part of the body, say a finger, and film of the operation by a worker is obtained with an illuminated bulb. Movement of the finger or hand is highlighted, which can be observed closely while viewing the film.

Chronocyclegraph

The bulb attached to the part of body is switched on and off frequently and the movement of the part is viewed as pear-shaped dots. By seeing the number of dots in a period or distance, one can decide whether the movement is fast or slow, or whether the part of the body is in acceleration or retardation.

    f. Man–machine chart: When an operator attends to one or more than one machine, then the interaction between man and machine is represented by the man–machine chart. It is also known as a multiple activity chart. After constructing the man–machine chart, one can determine:
      i. For how much time the machine is idle
     ii. For how much time the operator is idle

Consider one operator and one machine. The loading time of machine 1 is 0.2 min and it needs to run for 0.3 min in order to complete the operation. Assume that unloading is instantaneous. The time scale in minutes is shown in Figure 2.

    The operator is engaged for 0.2 min in loading the machine. For the next 0.3 min, the operator is idle while the machine is running. The cycle time is 0.5 min during which the machine is either being loaded or in actual operation.

    If a set of one operator and two machines is considered, then the man–machine chart is drawn as shown in Figure 3.

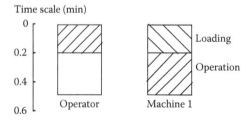

FIGURE 2 Man–machine chart for one operator and one machine.

Time scale (min)

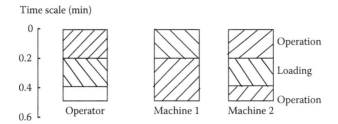

FIGURE 3 Man–machine chart for one operator and two machines.

Similar data are assumed for machine 2, also, i.e., loading time is 0.2 min and operation time is 0.3 min. In a cycle time of 0.5 min, the idle time of the operator is reduced to 0.1 min. For the first 0.2 min, the operator is loading machine 1, then for the next 0.2 min, the operator is loading machine 2, and idle for the remaining 0.1 min.

Example 5: An item requires two successive operations, which are planned to be done on two separate machines. The following data are given in minutes.

	Machine 1	Machine 2
Loading time	0.2	0.16
Operation time	0.2	0.21
Unloading time	0.1	0.1

Consider a set of one operator and two machines for the analysis and determine how many number of sets are required in order to produce 6,000 items per hour.

A man–machine chart is drawn as follows:

Time scale (min)

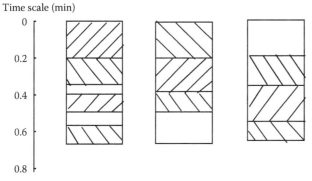

The cycle time obtained is 0.67 min. However, other sequences of activities should be worked out that may yield a lower cycle time.

One set may produce $60/0.67 = 89.55$ items per hour. The requirement is to produce 6,000 items per hour; therefore, the number of sets required $= 6,000/89.55 = 67$.

In other words, 67 operators and 134 machines (67 of each type) are needed to produce the desired output.

4.3 THERBLIGS

Seventeen elementary motions have been developed by Gilbreth that are known as "Therbligs". These are as follows.

S. No.	Therblig	Abbreviation
1	Select	SE
2	Use	U
3	Inspect	I
4	Transport empty	TE
5	Transport loaded	TL
6	Release load	RL
7	Pre-position	PP
8	Position	P
9	Disassemble	DA
10	Assemble	A
11	Plan	PN
12	Unavoidable delay	UD
13	Avoidable delay	AD
14	Rest to overcome fatigue	R
15	Search	S
16	Grasp	G
17	Hold	H

4.4 PRINCIPLES OF MOTION ECONOMY

All of the principles of motion economy have been categorized into three main categories:

A. Principles concerning the use of human body
B. Principles concerning the arrangement of workplace
C. Principles concerning tools and equipment design

Now, each one is briefly mentioned.

A. Principles concerning the use of human body

In order to minimize fatigue in the workplace, the human body must be used economically. The following parts of the body are arranged in the order of increasing fatigue while using them:

  i. Fingers
 ii. Fingers and wrist
iii. Fingers, wrist, and lower arm
 iv. Fingers, wrist, lower arm, upper arm, and movement of whole body

Rhythm needs to be generated at work, if possible, in order to enjoy the work and to decrease fatigue. Both hands may simultaneously start and stop the movement, if possible. Movement should follow smooth curves instead of changing the direction abruptly.

B. Principles concerning the arrangement of the workplace

A pleasant environment should always be in a workplace. Temperature, humidity, illumination, and ventilation should be at proper levels. Eyes should not be strained while working, including inspection of the components, etc.

Tools such as jigs/fixtures, etc. should be arranged in such a way that searching and selecting them becomes easier and takes less time. Clean surroundings always help in motion economy. Seating arrangements, such as chairs for workers, should be designed considering proper posture and comfort.

C. Principles concerning tools and equipment design

The design of tools and equipment should incorporate issues related to human comfort. For this purpose, ergonomics, i.e., human engineering, is applied. For instance, the handle of tools such as a screwdriver, needs to be long enough (at least equivalent to the width of the palm). This will ensure comfortable use.

Similarly, indicators of the equipment should be at the level of the operator's eyes, as far as possible, in order to facilitate efficient viewing. The arrangement of illumination may be made at the place in the equipment where frequent inspection is needed. The provision of a foot lever, if applicable, is useful from the point of view of motion economy.

4.5 Time Study

The objective of a time study is to evaluate the standard time of completing a job or element of an operation. It is also known as work measurement.

4.5.1 Applications

Various applications of time study or work measurement are as follows:

  a. After knowing the standard time for any job, it is easier to estimate the cost involved in the job. Time spent on any job can be converted into money spent

because people are paid on the basis of consumption of time, in addition to other factors.
b. Production planning becomes easier. With the help of standard time, production quantity in any period, say day/week, may be estimated.
c. It also helps in production control. If the job is not finished, an investigation is made and reasons are found for the deviation from the target.
d. Productivity of the workers/employees is increased if wage/salary determination has a bearing on the time study.

4.5.2   Procedure to Conduct Time Study

Steps to be followed for a time study are as follows:

a. The method of doing the job should be standardized before conducting time study. However, there may be an iterative procedure. Based on the time study, the method is improved so that it takes less time, and then the improved method is standardized, which may be used to conduct time study again.
b. The entire task is divided into smaller elements after a thorough study. This exercise is useful for recording the observed time of each element.
c. Average workers must be considered for conducting time study. If the selected workers are very fast, the observed time and eventually standard time computed will be less and most of the employees won't be able to complete the task in the standard time. This causes dissatisfaction among the employees. On the other hand, if workers selected for the time study are very slow, the standard time computed will be very high and productivity will be less. Therefore, workers selected should neither be very fast nor very slow. They should be normal workers.

A few selected workers are made to do the same task and the time taken is recorded with the help of a stopwatch. The average time is calculated, which may be called the observed time.

Now, normal time (basic time) = observed time × performance rating factor.

The performance rating factor (PRF), or rating factor, is used for leveling or normalizing the observed time. If an experienced industrial engineer feels that the observed time is too high, then the rating factor, or PRF, may be less than one. Similarly, if the observed time is too low, the PRF may be more than one.

The performance rating, or PRF, is the ratio of observed performance of the worker and normal performance from the industrial engineer's point of view.

The performance rating in % can be expressed as:

$$\text{Performance rating } (\%) = \frac{\text{Observed performance of the worker} \times 100}{\text{Normal performance}}$$

After obtaining the normal time, as discussed previously, the standard time is computed as follows:

$$\text{Standard time (S.T.)} = \text{Normal time (N.T.)} + \text{Allowances,}$$

where allowances are expressed in terms of % of normal (or basic) time.
The allowance are of the following types.

a. Setup allowance: Consider the following data.

Machine setup time = 60 min.
Once the machine is set up for a particular operation/process, it is ready for actual production. Normal time for production of one item on the machine is estimated as 7 min.

Due to the prevailing situations, the company is producing 60 items in one setup on that machine. The setup allowance for each item is the ratio of machine setup time and production volume in one setup, which is 60/60 = 1 min.

This allowance of 1 min can be expressed in terms of % of normal time as: $1/7 \times 100 = 14.29\%$.

This is also called a special allowance, which may also include other "one-time" activities before the actual production starts. Depending on the circumstances, the industry has to decide how to apportion this allowance on each item of production.

b. Allowance for unoccupied time: If there are interruptions in the work due to reasons such as power failures, then the allowance for unoccupied time should be included.

c. Interference allowance: If one operator is dealing with more than one machine, then one or more numbers of machines may wait for loading/unloading/starting. This idle time is accounted for as interference allowance.

d. Relaxation allowance: Workers cannot be engaged in the work throughout the day/shift continuously because of personal needs such as drinking water. This allowance accounts for time spent in all such activities.

e. Contingency allowance: There are certain actions that are difficult to predict. For example, the worker has to discuss some technical issues with the superior for some time. Time, thus elapsed, may be included in a contingency allowance.

Relevant allowances should be added to arrive at the value of total allowances.

Example 6: A job has been divided into four elements with the following data.

Element	Observed Time (min)	Performance Rating (%)
1	0.10	90
2	0.17	88
3	0.22	110
4	0.14	95

   i. Calculate the normal time for the job.
   ii. Find out the standard time for the job considering the following allowances as % of normal time:
      a. Setup allowance, 5%
      b. Relaxation allowance, 10%
      c. Interference allowance, 2%

Now:

   i. PRF (performance rating factor) for element 1 = 0.90

For element 2 = 0.88
   For element 3 = 1.10
   For element 4 = 0.95
   As the normal time = observed time × PRF.
   Normal time for element 1 = 0.1 × 0.90 = 0.09 min
   For element 2 = 0.17 × 0.88 = 0.1496 min
   For element 3 = 0.22 × 1.1 = 0.242 min
   For element 4 = 0.14 × 0.95 = 0.133 min
   Total normal time for the job = 0.6146 min.

   ii. Total allowances are 5 + 10 + 2 = 17% of the normal time.

   Value of allowances = 0.6146 × 0.17 = 0.104482 min.
   Standard Time = Normal Time + Allowances
   = 0.6146 + 0.104482
   = 0.719082 min
   ≈ 0.72 min.

4.5.3 Methods to Evaluate Performance Rating

As the performance rating is the ratio of observed performance and normal performance, the speed of the operator is judged with respect to a normal pace of working. This is called "speed rating".

In addition to speed rating, the following are the additional methods to evaluate the performance rating.

   a. Synthetic rating: Predetermined motion time standards (PMTS) have been developed for various elements. These standards are considered references. PRF is evaluated as the ratio of a predetermined standard for an element and observed time for that element as obtained in a time study at a workplace.
   b. Westinghouse system of rating: This is based on the following four factors:
      i. Skill
      ii. Effort
      iii. Conditions
      iv. Consistency

Every job requires a certain level of skill in performing it. The level of skill is measured. Efforts include physical as well as mental efforts in performing a job. The degree of effort involved is decided. Conditions refer to the working conditions and whether these are comfortable, hazardous, etc. Consistency relates to the level of performance with respect to time. It is observed whether performance (considering different parameters) is similar or deteriorates with respect to time.

In literature, these factors have been graded such as "excellent", "good", "average", "fair", etc. and the numbers are attached to each level.

Assume that for any task:

Skill corresponds to + 0.5

Effort corresponds to + 0.06

Conditions correspond to + 0.00

And consistency correspond to + 0.02.

Total = + 0.13.

PRF = 1 + 0.13 = 1.13.

In other words, the performance rating is 113%.

c. Objective rating: In addition to the usual speed rating, allowance is also included for the secondary adjustments, such as job difficulties.

For example, with the use of speed rating, the performance rating is 105%, i.e., 1.05%. Secondary adjustments are estimated as 125%, i.e., 1.25.

Finally, the PRF = 1.05 × 1.25 = 1.3125.

Exercises

1. What is single-machine scheduling? Describe the input parameters with respect to any job.
2. Explain the following:
   a. Completion time
   b. Flow time
   c. Queue time
   d. Lateness
   e. Tardiness
   f. Mean number of jobs in process
3. What are the various measures of performance in the context of sequencing of jobs with a single machine?
4. Name the measures of performance that are minimized, using:
   a. SPT rule
   b. EDD order
5. The following six jobs are ready at the same time, i.e., at time zero, for processing on a single machine. Apply the shortest processing time (SPT) rule:

J:	1	2	3	4	5	6
D_j	12	9	5	23	7	6
t_j	6	3	4	13	2	5

Also, evaluate the measures of performance that will be minimized.
6. The following seven jobs are ready at time zero for operations on a single machine:

J:	1	2	3	4	5	6	7
D_j	15	7	6	20	5	16	9
t_j	9	2	3	15	4	9	8

Sequence the jobs in EDD order and evaluate the performance measures that will be minimized using EDD.
7. Explain the method to minimize the number of tardy jobs and apply it for the following data:

J:	1	2	3	4	5	6	7
D_j	18	10	9	10	4	14	8
t_j	13	6	7	17	2	10	6

8. (a) Write the steps involved in Johnson's algorithm.
   (b) Consider a two-machine flow shop. The operation time of jobs on each machine are as follows:

Job	Time	
	Machine 1	Machine 2
J_1	7	6
J_2	8	9
J_3	3	9
J_4	13	11
J_5	5	10

Sequence the jobs so that the make span is minimized. Also, construct the Gantt chart and comment on it.

9. What is productivity?
10. How is work study useful in increasing productivity?
11. Discuss the procedure for conducting a method study.
12. What are the basic symbols used in a motion study?
13. Explain the procedure of creating a flow process chart.
14. In which situation is a string diagram more useful?
15. What is a micromotion study? Discuss the following:
    a. Cyclegraph
    b. Chronocyclegraph

BIBLIOGRAPHY

1. Chase R.B., Aquilano N.J., and Jacobs F.R. *Production and Operations Management.* TMH, 2000.
2. Nahmias S. *Production and Operations Analysis.* McGraw-Hill, 2001.
3. Panneerselvam R. *Production and Operations Management.* PHI, 1999.
4. Smith S.B. *Computer-Based Production and Inventory Control.* Prentice-Hall, 1989.

Index

CPSIA information can be obtained
at www.ICGtesting.com
Printed in the USA
JSHW020221110522
25519JS00001BA/34